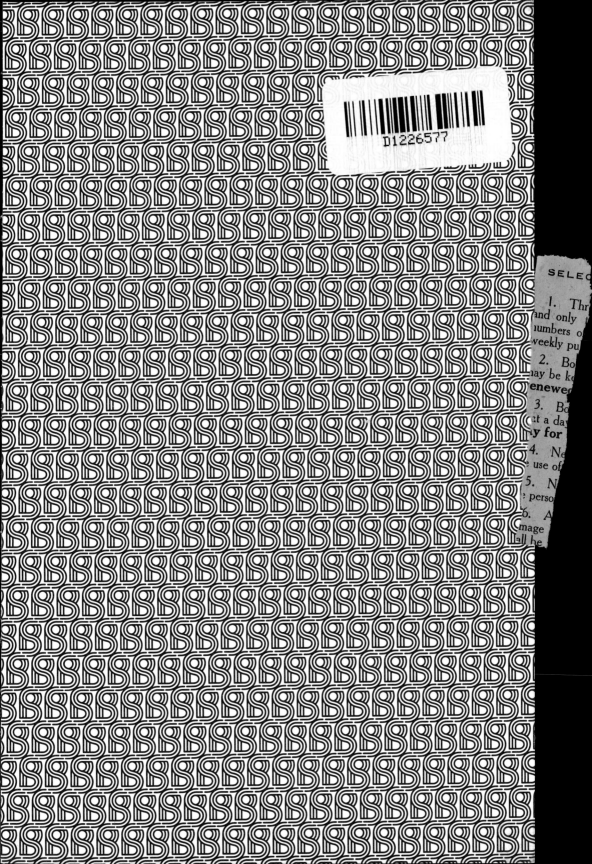

m
r

ce
da

the

the

da
sh

Stripe
Press

Ideas for progress
San Francisco, California
press.stripe.com

An Elegant Puzzle

SYSTEMS OF ENGINEERING MANAGEMENT

Summary

There's a saying that people don't leave companies, they leave managers. Management is a key part of any organization, yet the discipline is often self-taught and unstructured. Getting to good solutions for complex management challenges can make the difference between fulfillment and frustration for teams, and, ultimately, between the success or failure of companies.

Will Larson's *An Elegant Puzzle* orients around the particular challenges of engineering management—from sizing teams to managing technical debt to succession planning—and provides a path to the good solutions. Drawing from his experience at Digg, Uber, and Stripe, Will Larson has developed a thoughtful approach to engineering management that leaders of all levels at companies of all sizes can apply. *An Elegant Puzzle* balances structured principles and human-centric thinking to help any leader create more effective and rewarding organizations for engineers to thrive in.

Bio

Will Larson has been an engineering leader and software engineer at technology companies of many shapes and sizes, including Yahoo!, Digg, SocialCode, Uber, and, since 2016, Stripe.

He grew up in North Carolina, studied Computer Science at Centre College in Kentucky, spent a year in Japan for the JET Program teaching English, and has been living in San Francisco since 2009.

An Elegant Puzzle draws from the writing in his blog, *Irrational Exuberance!,* that he has been updating since graduating from college. It is currently, and will always be, a work in progress.

@lethain
www.lethain.com

An Elegant Puzzle: Systems of Engineering Management

First published in 2019 in hardcover
in the United States of America
by Stripe Press / Stripe Matter Inc.

Stripe Press
Ideas for progress
San Francisco, California
press.stripe.com

Printed in Canada
ISBN: 978-1-7322651-8-9

First Edition

Contents

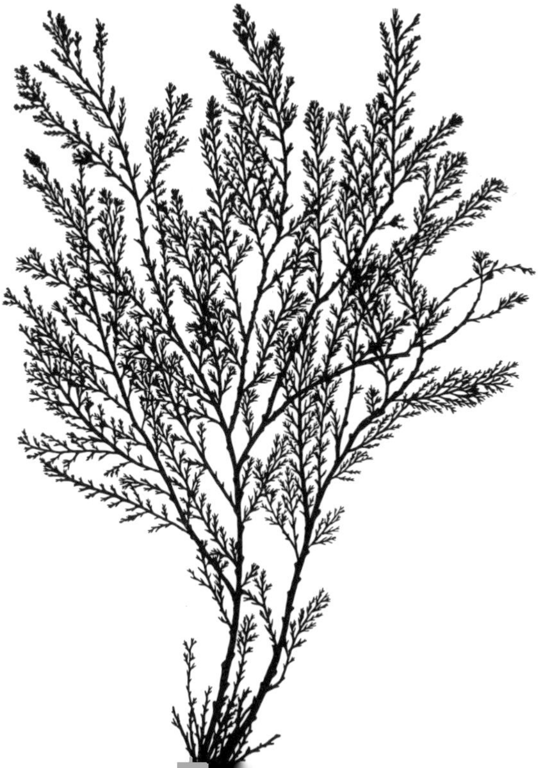

Preface

The first blog post that I ever wrote was on April 7, 2007, and was titled "Finding Our Programming Flow." It was not very good. That year I wrote 69 posts, the last being "Miyajima and Hiroshima," a collection of pictures from a trip I took while teaching English in Japan. The next year, 2008, I wrote 192 posts. The writing still left much to be desired.

It took 200 more posts and another decade to cobble together a written voice and to make enough mistakes that my experience might become worth reading. I'm fortunate that that moment coincided with my time at Stripe, an environment where folks routinely do things that might in other contexts seem out of reach: things like starting a technology magazine or publishing a book.

This book is the lucky alignment of that happy moment at Stripe, a decade of writing and a decade of learning about both leadership and management, and the good fortune to work with colleagues who suggested collecting my writing into this book.

I hope that everyone who picks up this book gets something useful out of it. Your comments and thoughts are appreciated at:
lethain@gmail.com.

Acknowledgments

A few particular thanks are in order. This book came together in the months preceding my wedding, and I'm grateful to Laurel for her thoughts and partnership. Thank you to Brianna Wolfson and Tyler Thompson, without whom this book would not exist. Finally, thanks to my sister, Hope, who has long been the talented sibling, showing that the path to authorship is attainable.

An Elegant Puzzle

SYSTEMS OF ENGINEERING MANAGEMENT

WILL LARSON

1

CHAPTER 1

Introduction

1.1 Introduction

Some people go into management out of a desire to be of service. Others become managers in a cynical pact, exchanging excitement in their current role for the prospect of continued salary bumps and promotions. There are even folks who initially go into management because they're entirely fed up with their own manager and are convinced that they could do better.

I won't say which of those, if any, describes me.

Regardless of what motivation first brings you into management, it can feel as if you've entered a troubled profession. Skilled practitioners are scarce, and only the exceptional company is willing to invest in growing its managers.

If training programs are peculiarly uncommon, concerns that today's managers are ill-prepared are not. I was lucky early in my management career to have a coworker describe me as the best leader they had worked with. It took several additional years of practice for another to declare me their worst.

While David's cloak is increasingly suspicious draped from the shoulders of Silicon Valley's Goliaths, the vast majority of technology companies are well-meaning chrysalises that hope to one day birth a successful business, and they are guided by managers who are learning to lead, one unexpected lesson at a time. For many such people, the entry into engineering management begins with a crisis, and their training is a series of hard knocks.

This was certainly my experience: my path into management began at Digg, paved by a pair of layoffs in 2010. The three one-on-ones I'd had at my previous job had not, surprisingly, culminated in a rigorous framework for management, and I had absolutely no idea what I was doing.

In the years since, I've worked to educate myself on the topic, reading anything that seemed even distantly relevant. There are some wonderful resources out there (many of which I've listed in the "Books I've found very useful" appendix), but the rare answers I uncovered continued to be drowned out by my ever-increasing sea of questions.

It was only when I got the opportunity to work at Uber, which was growing its engineering team from 200 to 2,000 over two years, and then at Stripe, which was experiencing similar rapid growth, that I had to opportunity to truly refine my approach to management through exposure to an endless variety of challenges. There are few things peaceful about managing in rapidly growing companies, but I've never found anywhere better to learn and to grow.

As I've become more experienced, my appreciation for management, and engineering management in particular, has grown, and I've come to view the field as a series of elegant, rewarding, and important puzzles. This book is a collection of those puzzles, which I've had the good fortune to struggle with and learn from. It starts with the most important tool in my kit, "Organizations." Organizational design gets the right people in the right places, empowers them to make decisions, and then holds them accountable for their results. Maintained consistently and changed sparingly, nothing else will help you scale more. Next, we'll review a handful of fundamental "Tools" of management that I've found useful across a wide variety of scenarios. These range from systems thinking to vision documents, from metrics to migrations, from reorgs to career narratives. Perhaps the easiest way to use this chapter is to skim over the ideas quickly, and then reread them when it seems as if they might be useful.

The third chapter, "Approaches," works through circumstances in which you might need to adjust how you manage. It digs into how to adapt your management for rapidly growing companies, and how to manage when your desired impact is beyond your authority. It will hopefully help you find alternative paths to approach some of the areas where you're not feeling as successful as you'd like.

That chapter is followed by an exploration of "Culture." This section is focused on practical things you can do to nurture an inclusive team or organization. It also digs into a few particular culture topics on the conflict of "freedom to" and "freedom from," and deals with hero culture.

Finally, the book ends with an exploration of "Careers," with a focus on interviewing, hiring, and performance management. Many managers come up thinking of recruiting as something run by recruiters, and performance management as something designed by human resources, but these are powerful tools that you should be using frequently.

If you finish the entire book, you won't walk into your office the next day as a perfect manager (I remain grateful for the days I walk into the office feeling like a marginally competent one), but I hope that it'll stimulate questions about how you're approaching management, provide a few new approaches for you to experiment with, and help you take a few steps further down the path of engineering management.

2

CHAPTER 2

Organizations

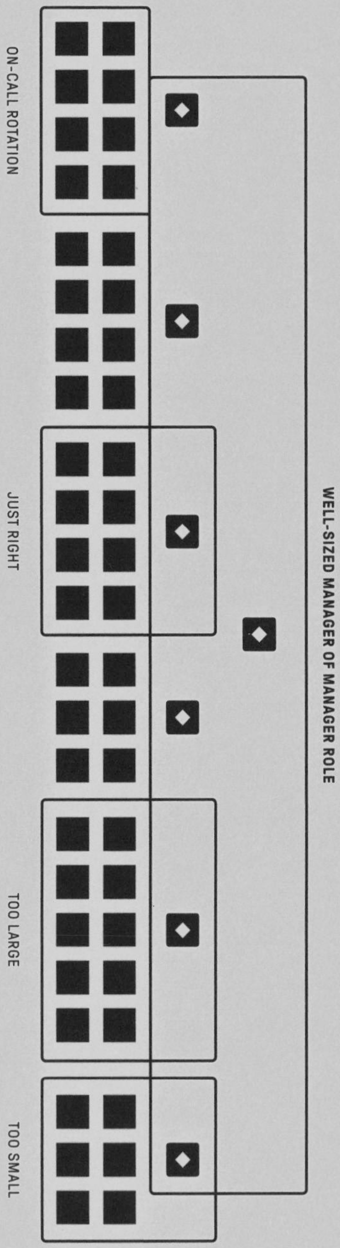

Figure 2.1
Sizing teams and groups of teams using sizing rules.

Organizations

An organization is a collection of people working toward a shared goal. Each organization is an exploration of the possible, undertaken together by the ten, the hundred, or the thousand. Initially, I was tempted to glibly write that sometimes organizations work, but the truly extraordinary thing is that all organizations work.

Some do indeed work better than others, and organizational design is the attempt to understand why some create such energy and others create mostly heat: friction, frustration, and politics. I believe that excellent organizations grow from consistently applying a straightforward process.

When I have a problem that I want to solve quickly and cheaply, I start thinking about process design. A problem I want to solve permanently and we have time to go slow? That's a good time to evolve your culture. However, if process is too weak a force, and culture too slow, then organizational design lives between those two.

This chapter covers the approaches to organizational design and evolution that I've found effective. If you're reading through and find yourself thinking that this sounds easy, I agree! The hard bit is keeping your courage up when circumstances get challenging.

2.1 Sizing teams

When I transitioned from supporting a team to supporting an organization, I started to encounter a new category of problems that I had never thought about. How many teams should we have? Should we create a new team for this initiative, or ask an existing team to take it on? What is the boundary between these two teams?

These questions were the gateway to the obscure art of organizational design. As I've gotten more exposure, I've come to believe that the fundamental challenge of organizational design is sizing teams. You'll find yourself sizing teams during reorganizations,[1] to accommodate growth from hiring, and when considering how to support new projects. It'll be an unusual month that you won't consider some aspect of team design.

While I'm skeptical that there exists a unified law of team sizing, I have iterated my requirements onto a useful framework that solves the majority of cases I encounter. That framework has in turn led to a standard playbook. Both are short, opinionated, and hopefully useful!

The guiding principles I use for sizing teams are:

⊙ Managers should support six to eight engineers

This gives them enough time for active coaching, coordinating, and furthering their team's mission by writing strategies,[2] leading change,[3] and so on.

Tech Lead Managers (TLMs). Managers supporting fewer than four engineers tend to function as TLMs, taking on a share of design and implementation work. For some folks this role can uniquely leverage their strengths, but it's a role with limited career opportunities. To progress as a manager, they'll want more time to focus on developing their management skills. Alternatively, to progress toward staff engineering roles, they'll find it difficult to spend enough time on the technical details.

Coaches. Managers supporting more than eight or nine engineers typically act as coaches and safety nets for problems. They are too busy to actively invest in their team or their team's area of responsibility. It's reasonable to ask managers to support larger teams during the transition to a more stable configuration, but it is a bad status quo.

⊙ Managers-of-managers should support four to six managers

This gives them enough time to coach, to align with stakeholders, and to do a reasonable amount of investment in their organization. On the other hand, it will also keep them busy enough that they won't be tempted to create work for their team.

Ramping up. Managers supporting fewer than four other managers should be in a period of active learning on either the problem domain or on transitioning from supporting engineers to supporting managers. In the steady state, this can lead to folks feeling underutilized, or being tempted to meddle in daily operations.

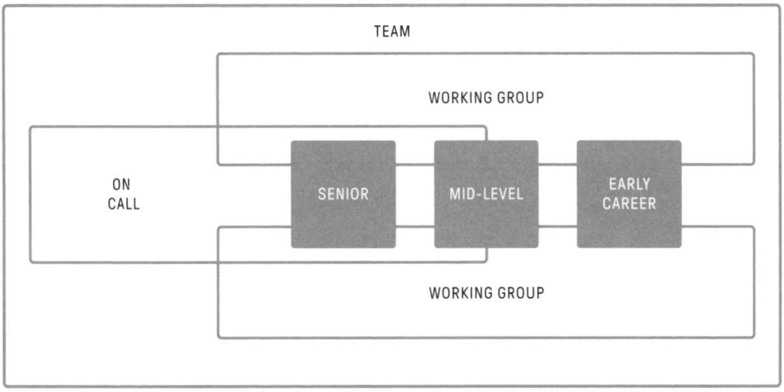

Figure 2.2
A team composed of two working groups, an on-call rotation, and different tenured engineers.

Coaches. Similar to supporting a large team of engineers, supporting a large team of managers leaves you functioning purely as a problem-solving coach.

On-call rotations want eight engineers

For production on-call responsibilities,[4] I've found that two-tier 24/7 support requires eight engineers. As teams holding their own pagers have become increasingly mainstream, this has become an important sizing constraint, and I try to ensure that every engineering team's steady state is eight people.

Shared rotations. It is sometimes necessary to pool multiple teams together to reach the eight engineers necessary for a 24/7 on-call rotation. This is an effective intermediate step toward teams owning their own on-call rotations, but it is not a good long-term solution. Most folks find being on-call for components that they're unfamiliar with to be disproportionately stressful.

Small teams (fewer than four members) are not teams

I've sponsored quite a few teams of one or two people, and each time I've regretted it. To repeat: I have regretted it every single time. An important property of teams is that they abstract the complexities of the individuals that compose them. Teams with fewer than four individuals are a sufficiently leaky abstraction that they function indistinguishably from individuals. To reason about a small team's delivery, you'll have to know about each on-call shift, vacation, and interruption.

They are also fragile, with one departure easily moving them from innovation back into toiling to maintain technical debt.

Keep innovation and maintenance together. A frequent practice is to spin up a new team to innovate while existing teams are bogged down in maintenance. I've historically done this myself, but I've moved toward innovating within existing teams.[5] This requires very deliberate decision-making and some bravery, but in exchange you'll get higher morale and a culture of learning, and will avoid creating a two-tiered class system of innovators and maintainers.

Fitting together those guiding principles, the playbook that I've developed is surprisingly simple and effective:

- Teams should be six to eight during steady state.
- To create a new team, grow an existing team to eight to ten, and then bud into two teams of four or five.
- Never create empty teams.
- Never leave managers supporting more than eight individuals.

Like all guidelines, this is a structure to aid thinking through sizing problems, not a straitjacket to restrict every exception. The context of any situation deserves careful examination, but increasingly I've found that the long-term costs of exceptions outweigh what I once considered their strengths.

2.2 Staying on the path to high-performing teams

A friend is six months into supporting a 60 person engineering group. Perhaps unsurprisingly, most of their teams believe that they have urgent hiring needs. Should my friend spread hiring equally across the teams in need, or focus hiring on just one or two teams until their needs are fully staffed? That is the question.

It's a great question, and captures a deeply challenging aspect of leading an organization. It's fun to do initial discovery, learning from and about everyone. The rare moment when you choose to reorganize the team[6] is painful, but concludes quickly. What's much harder is keeping the faith when you've played your cards and need to find space for your plans to come to fruition. Staying the course is particularly fraught when it comes to growing an organization, because some teams always need more than you choose to provide.

When you talk about growing an organization, the conversation usually leads to hiring. While I believe that hiring is a very important approach to growing organizations, I also believe that we reach for it too often. In order to prioritize hiring for scenarios in which it'll do the most good, over the past year I've developed a loose framework for reasoning about what a given team needs to increase performance.

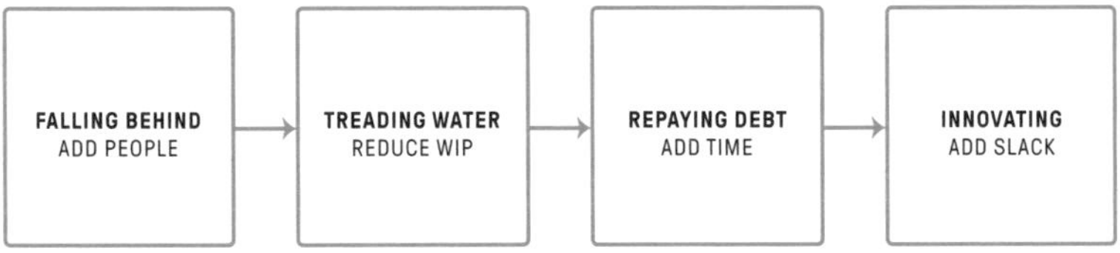

Figure 2.3
Four states of a team.

2.2.1 Four states of a team

The framework starts with a vocabulary for describing teams and their performance within their surrounding context.

Teams are slotted into a continuum of four states:

A team is falling behind if each week their backlog is longer than it was the week before. Typically, people are working extremely hard but not making much progress, morale is low, and your users are vocally dissatisfied.

A team is treading water if they're able to get their critical work done, but are not able to start paying down technical debt or begin major new projects. Morale is a bit higher, but people are still working hard, and your users may seem happier because they've learned that asking for help won't go anywhere.

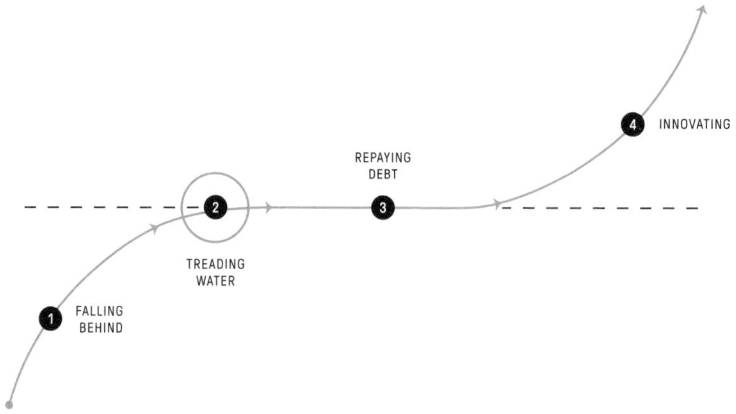

Figure 2.4
Four stages of a team's progress, from falling behind to innovating.

A team is repaying debt when they're able to start paying down technical debt, and are beginning to benefit from the debt repayment snowball: each piece of debt you repay leads to more time to repay more debt.

A team is innovating when their technical debt is sustainably low, morale is high, and the majority of work is satisfying new user needs.

Teams want to climb from *falling behind* to *innovating*, while entropy drags them backward. Each state requires a different tact.

2.2.2 System fixes and tactical support

In this framework, teams transition to a new state exclusively by adopting the appropriate **system solution** for their current state. As a manager, your obligation is to identify the correct system solution for a given transition, initiate that solution, and then support the team as best you can to create space for the solutions to work their magic. If you skip to supporting the team tactically before initiating the correct system solution, you'll exhaust yourself with no promise of salvation. For each state, here is the strategic solution that I've found most effective, along with some ideas about how to support the team while that solution comes to fruition:

1. **When the team is falling behind**, the system fix is to hire more people until the team moves into treading water. Provide tactical support by setting expectations with users, beating the drum around the easy wins you can find, and injecting optimism.

   As a caveat, the system fix is to hire net new people, increasing the overall capacity of the company. Sometimes people instead attempt to capture more resources from the existing company, and I'm pretty negative on that. People are not fungible, and generally folks end up in useful places, so I'm skeptical of reassigning existing individuals to drive optimality. By nature, it's also impossible for this kind of discussion to not become political, even when everyone involved has deep trust in and respect for each other.

2. **When the team is treading water**, the system fix is to consolidate the team's efforts to finish more things, and to reduce concurrent work until they're able to begin repaying debt (e.g., limit work in progress). Tactically, the focus here is on helping people transition from a personal view of productivity to a team view.

3. **When the team is repaying debt**, the system fix is to add time. Everything is already working, you just need to find space to allow the compounding value of paying down technical debt to grow. Tactically try to find ways to support your users while also repaying debt, to avoid disappearing into technical debt repayment from your users' perspective. Especially for a team that started out falling behind and is now repaying debt, your stakeholders are probably antsy waiting for the team to start delivering new stuff, and your obligation is to prevent that impatience from causing a backslide!

4. **Innovating** is a bit different, because you've nominally reached the end of the continuum, but there is still a system fix! In this case, it's to maintain enough slack in your team's schedule that the team can build quality into their work, operate continuously in innovation, and avoid backtracking. Tactically, ensure that the work your team is doing is valued: the quickest path out of innovation is to be viewed as a team that builds science projects, which inevitably leads to the team being defunded.

   I can't stress enough that these fixes are *slow*. This is because systems accumulate months or years of static, and you have to drain that all away. Conversely, the same properties that make these fixes slow to fix make them extremely durable once in effect!

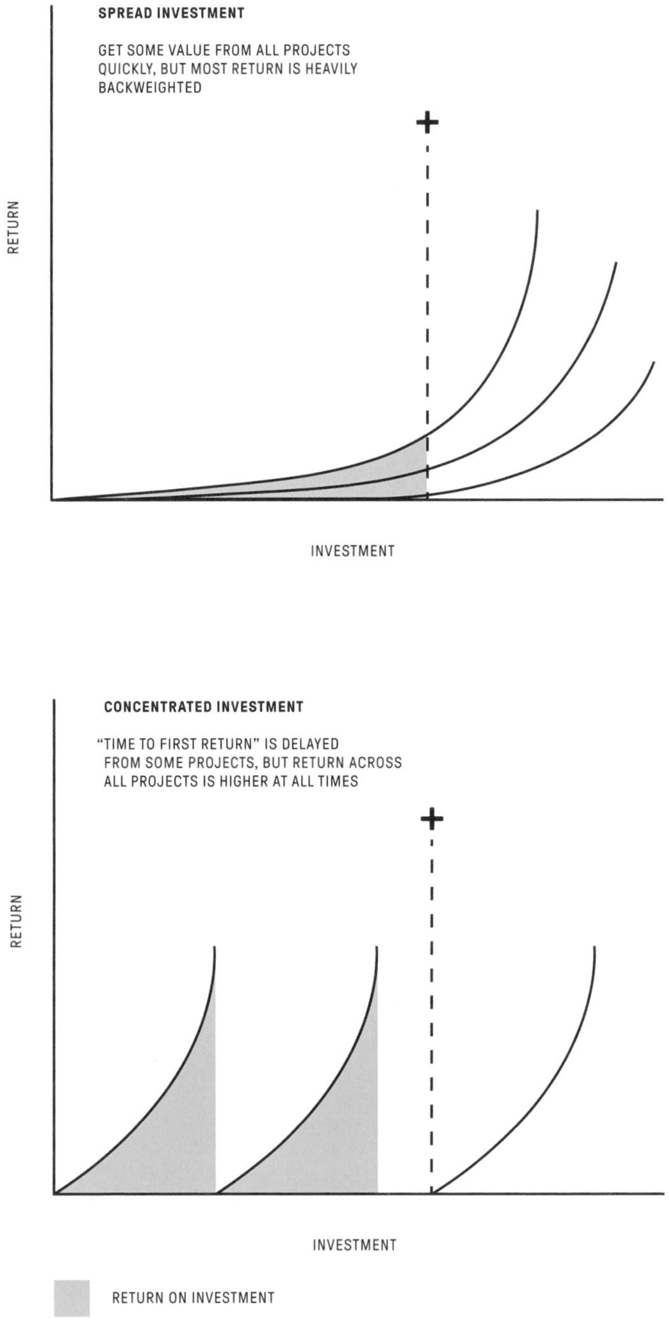
SPREAD INVESTMENT
GET SOME VALUE FROM ALL PROJECTS QUICKLY, BUT MOST RETURN IS HEAVILY BACKWEIGHTED
RETURN
INVESTMENT
CONCENTRATED INVESTMENT
"TIME TO FIRST RETURN" IS DELAYED FROM SOME PROJECTS, BUT RETURN ACROSS ALL PROJECTS IS HIGHER AT ALL TIMES
RETURN
INVESTMENT
RETURN ON INVESTMENT

Figure 2.5
Return on investment when consolidating versus spreading investment.

The hard part is maintaining faith in your plan—both your faith and the broader organization's faith. At some point, you may want to launder accountability through a reorg, or maybe skip out to a new job, but if you do that you're also skipping the part where you get to learn. Stay the path.

2.2.3 Consolidate your efforts

As an organizational leader, you'll be dealing with a number of teams, each of which is in a different place on this continuum. You'll also have limited resources to apply, and they'll usually be insufficient to simultaneously move every team down the continuum. Many folks try to move all teams at the same time, peanut buttering[7] their limited resources, but resist that indecision-framed-as-fairness: it's not a fair outcome if no one gets anything.

For each constraint, prioritize one team at a time. If most teams are falling behind, then hire onto one team until it's staffed enough to tread water, and only then move to the next. While this is true for all constraints, it's particularly important for hiring.

Adding new individuals to a team disrupts that team's gelling process, so I've found it much easier to have rapid growth periods for any given team, followed by consolidation/gelling periods during which the team gels. The organization will never stop growing, but each team will.

2.2.4 Durable excellence

This approach to nurturing great organizations is the opposite of a quick fix. While it's slow, I've found that it consistently leads to enduring, real improvement in the happiness and throughput of an organization. Most importantly, these improvements stick around long enough to compound, creating a durable excellence.

2.3 A case against top-down global optimization

After I wrote "Staying on the Path to High-Performance Teams,"[8] quite a few people asked the same follow-up question: "Once a team has repaid its technical debt, shouldn't the now surplus team members move to other teams?"

This makes a lot of sense, because the team, with so little technical debt left, is now overstaffed relative to its global priority. Repeated across many teams, this could lead to an organization having far too many engineers allocated against last year's problems, and too few against today's.

This is an important problem to address!

First, let me explain why I'm skeptical of reallocating individuals to address global priority shifts, and then I'll suggest a couple alternative approaches to this conundrum.

2.3.1 Team first

Fundamentally, I believe that sustained productivity comes from high-performing teams, and that disassembling a high-performing team leads to a significant loss of productivity, even if the members are fully retained. In this worldview, high-performing teams are sacred, and I'm quite hesitant to disassemble them.

Teams take a long time to gel. When a group has been working together for a few years, they understand each other and know how to set each other up for success in a truly remarkable way. Shifting individuals across teams can reset the clock on gelling, especially for teams in the early stages of gelling, and when there are significant differences in team culture. That's not to say that you want teams to never change—that leads to stagnation—but perhaps preserving a team's gelled state requires restrained growth.

Sometimes you will want to grow faster than a gelled team allows, and that's okay! The lesson is that you have to account for re-gelling costs after periods of change, not that you should never change them. This is part of why my proposed model[9] recommends rapidly hiring into teams loaded down by technical debt, not into innovating teams, which avoids incurring re-gelling costs on high-performing teams.

2.3.2 Fixed costs

Another reason that I lean away from moving folks off high-performing teams is that most teams have high fixed costs and relatively small variable costs: moving one person can shift an innovating team back into falling behind, and now neither team is doing particularly well. This is especially true on teams responsible for products and services.

My rule of thumb is that it takes eight engineers on a team to support a two-tier on-call rotation, so I'm generally reluctant to move any team with membership below that line. However, fixed costs come in many other varieties: "keeping the lights on" work, precommitted contracts, support questions from other teams, etc.

There are some teams with very low fixed costs—a startup without any users, a team supporting a product that you've turned off entirely—and I suspect that the rules for those teams are different. I also suspect that such teams are quite uncommon in successful companies.

2.3.3 Slack

The premise of moving folks to optimize global efficiency also implies a deeper understanding of how productivity is generated than I've ever personally achieved. I'm a strong believer in not adding more resources to a team with visible slack, but I'm less convinced that the inverse applies.

The expected time to complete a new task approaches infinity as a team's utilization approaches 100 percent, and most teams have many dependencies on other teams. Together, these facts mean you can often slow a team down by shifting resources to it, because doing so creates new upstream constraints.

In further defense of slack, I find that teams put spare capacity to great use by improving areas within their aegis, in both incremental and novel ways. As a bonus, they tend to do these improvements with minimal coordination costs, such that the local productivity doesn't introduce drag on the surrounding system.

Most importantly, "slackful" teams function as an organizational debugger: you don't have to consider them when debugging the overall organizational throughput. I've found it much easier to work a couple constraints at a time, solving forward without needing to revisit previous constraints.

The Goal by Eliyahu M. Goldratt[10] and *Thinking in Systems: A Primer* by Donella H. Meadows[11] are both phenomenal books on this topic.

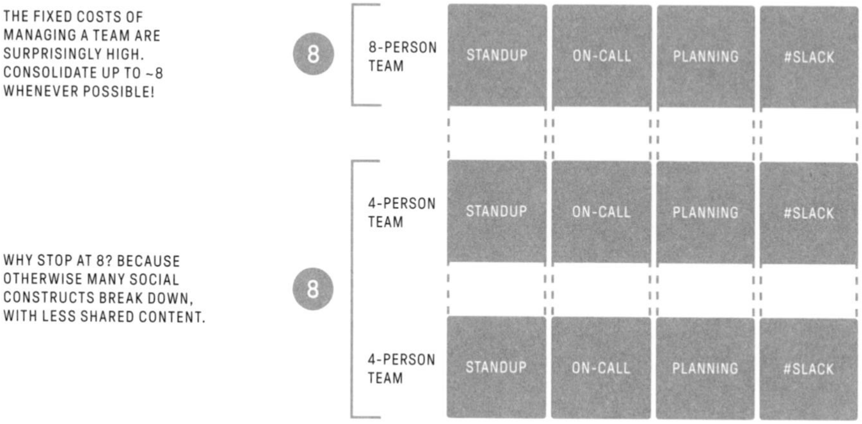
THE FIXED COSTS OF MANAGING A TEAM ARE SURPRISINGLY HIGH. CONSOLIDATE UP TO ~8 WHENEVER POSSIBLE!
WHY STOP AT 8? BECAUSE OTHERWISE MANY SOCIAL CONSTRUCTS BREAK DOWN, WITH LESS SHARED CONTENT.
8
8-PERSON TEAM
STANDUP
ON-CALL
PLANNING
#SLACK
8
4-PERSON TEAM
STANDUP
ON-CALL
PLANNING
#SLACK
4-PERSON TEAM
STANDUP
ON-CALL
PLANNING
#SLACK

Figure 2.6
Fixed costs of running a team.

2.3.4 Shift scope; rotate

Okay, so what *does* work? I've found it most fruitful to move scope between teams, preserving the teams themselves. If a team has significant slack, then incrementally move responsibility to them, at which point they'll start locally optimizing their expanded workload. It's best to do this slowly to maintain slack in the team, but if it's a choice of moving people rapidly or shifting scope rapidly, I've found that the latter is more effective and less disruptive.

Shifting scope works better than moving people because it avoids re-gelling costs, and it preserves system behavior. Preserving behavior keeps your existing mental model intact, and if it doesn't work out, you can always revert a workload change with less disruption than would be caused by a staffing change.

The other approach that I've seen work well is to rotate individuals for a fixed period into an area that needs help. The fixed duration allows them to retain their identity and membership in their current team, giving their full focus to helping out, rather than splitting their focus between performing the work and finding membership in the new team. This is also a safe way to measure how much slack the team really has!

A coworker of mine suggested that some companies have very successfully moved toward the swarming model (at the organization level, not just at the team level), and I hope that I eventually get a chance to hear from people who've successfully gone the other direction! One of the most exciting aspects of organizational design is that there are so many different approaches that work well.

2.4 Productivity in the age of hypergrowth

You don't hear the term hypergrowth[12] quite as much as you did a couple years ago. Sure, you might hear it during any given week, but you also might open up Techmeme and not see it, which is a monumental return to a kinder, gentler past. (Or perhaps we're just unicorning[13] now.)

Fortunately for engineering managers everywhere, the challenges of managing within quickly growing companies still very much exist.

When I started at Uber, we were almost 1,000 employees and were doubling the headcount every six months. An old-timer summarized their experience as: "We're growing so quickly that every six months we're a new company." A bystander quickly added a corollary: "Which means our process is always six months behind our head count."

Helping my team be successful when a defunct process merges with a constant influx of new engineers and system load has been one of the most rewarding opportunities I've had in my career. This is an attempt to explore the challenges and propose some strategies I've seen for mitigating and overcoming them.

2.4.1 More engineers, more problems

All real-world systems have some degree of inherent self-healing properties: an overloaded database will slow down enough that someone fixes it, and overwhelmed employees will get slow at finishing work until someone finds a way to help.

Very few real-world systems have efficient and deliberate self-healing properties, and this is where things get exciting as you double engineers and customers year after year after year.

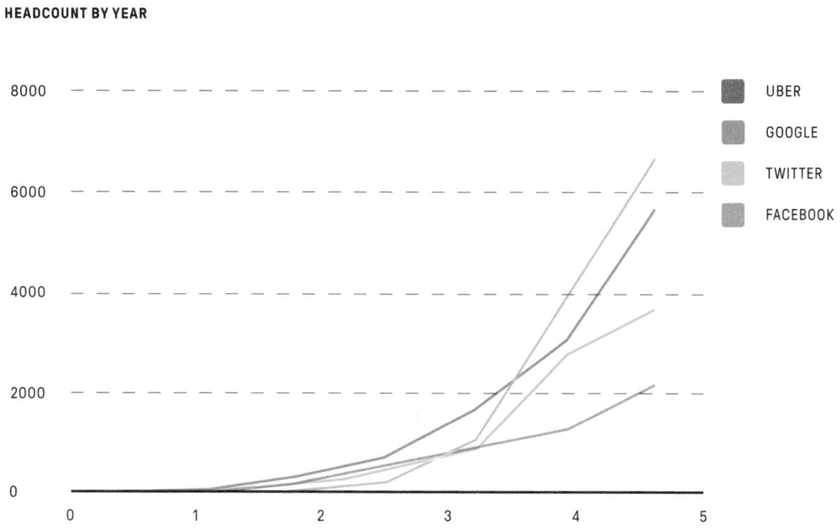

Figure 2.7
Employee growth rate of fast-growing companies.

Productively integrating large numbers of engineers is hard.

Just how challenging this is depends on how quickly you can ramp engineers up to self-sufficient productivity, but if you're doubling every six months and it takes six to twelve months to ramp up, then you can quickly find a scenario in which untrained engineers increasingly outnumber the trained engineers, and each trained engineer is devoting much of their time to training a couple of newer engineers.

Imagine a scenario in which training a single new engineer takes about 10 hours per week from each trained engineer, and in which untrained engineers are one-third as productive as trained engineers. The result is the right-hand chart's (admittedly, pretty worst-case scenario) ratio of two-untrained-to-one-trained. Worse, for those three people you're only getting the productivity of 1.16 trained engineers (2 x .33 for the untrained engineers plus .5 x 1 for the trainer).

You also need to factor in the time spent on hiring.

If you're trying to double every six months, and about 10 percent of candidates undergoing phone screens eventually join, then you need to do ten interviews per existing engineer in that time period, with each interview taking about two hours to prep, perform, and debrief.

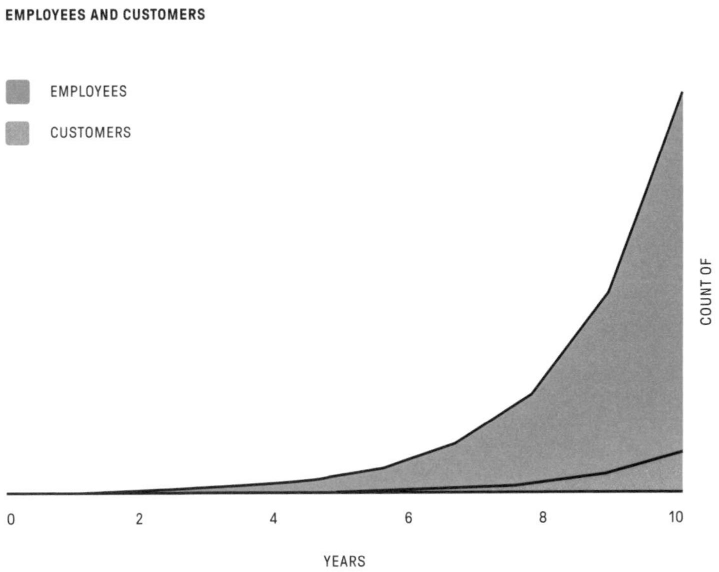

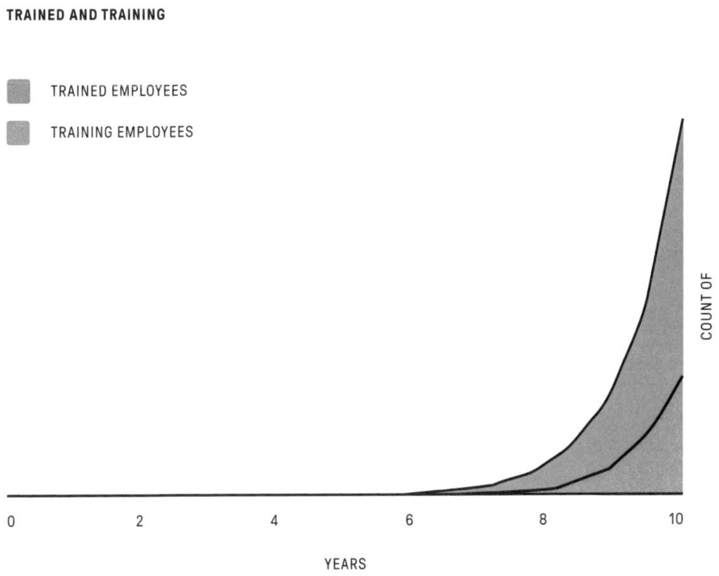

Figure 2.8
More employees, more customers, more problems.

That's less than four hours per engineer per month if you can leverage your entire existing team, but training comes up again here: if it takes you six months to get the average engineer onto your interview loop, each trained engineer is now doing three to four hours of hiring-related work per week, and your trained engineers are down to approximately 0.4 efficiency. The overall team is getting 1.06 engineers' worth of work out of every three engineers.

It's not just training and hiring, though:

1. For every additional order of magnitude of engineers, you need to design and maintain a new layer of management.

2. For every ~10 engineers, you need an additional team, which requires more coordination.[14]

3. Each engineer means more commits and deployments per day, creating load on your development tools.

4. Most outages are caused by deployments, so more deployments drive more outages, which in turn require incident management, mitigations, and postmortems.

5. Having more engineers leads to more specialized teams and systems, which require increasingly small on-call rotations so that your on-call engineers have enough system context to debug and resolve production issues. Consequently, relative time invested in on-call goes up.

Let's do a bit more handwavy math to factor these in.

Only your trained engineers can go on-call. They're on-call one week a month, and are busy about half their time on-call. So that's a total impact of five hours per week for your trained engineers, who are now down to 0.275 efficiency, and your team is now getting less than the output of a single trained engineer for every three engineers you've hired.

(This is admittedly an unfair comparison because it's not accounting for the on-call load on the smaller initial teams, but if you accept the premise that on-call load grows as engineer head count grows, and that load grows as the number of rotations grows, then the conclusion should still roughly hold.)

Although it's rarely quite this extreme, this is where the oft-raised concern that "hiring is slowing us down" comes from: at high enough rates, the marginal added value of hiring gets very slow, especially if your training process is weak.

Sometimes very low means negative!

2.4.2 Systems survive one magnitude of growth

We've looked a bit at productivity's tortured relationship with engineering head count, so now let's also think a bit about how the load on your systems is growing.

Understanding the overall impact of increased load comes down to a few important trends:

1. Most system-implemented systems are designed to support one to two orders' magnitude of growth from the current load. Even systems designed for more growth tend to run into limitations within one to two orders of magnitude.

2. If your traffic doubles every six months, then your load increases an order of magnitude every 18 months. (And sometimes new features or products cause load to increase much more quickly.)

3. The cardinality of supported systems increases over time as you add teams, and as "trivial" systems go from unsupported afterthoughts to focal points for entire teams as the systems reach scaling plateaus (things like Apache Kafka, mail delivery, Redis, etc.).

If your company is designing systems to last one order of magnitude and is doubling every six months, then you'll have to re-implement every system twice every three years. This creates a great deal of risk—almost every platform team is working on a critical scaling project—and can also create a great deal of resource contention to finish these concurrent rewrites.

However, the real productivity killer is not system rewrites but the migrations that follow those rewrites. Poorly designed migrations expand the consequences of this rewrite loop from the individual teams supporting the systems to the entire surrounding organization.

If each migration takes a week, each team is eight engineers, and you're doing four migrations a year, then you're losing about 1 percent of your company's total productivity. If each of those migrations takes closer to a month, or if they are only possible for your small cadre of trained engineers—whose time is already tightly contended for—then the impact becomes far more pronounced.

There is a lot more that could be said here—companies that mature rapidly often have tight and urgent deadlines around pursuing various critical projects, and around moving to multiple data centers, to active-active designs, and to new international regions—but I think we've covered our bases on how increasing system load can become a drag on overall engineering throughput.

The real question is, what do we do about any of this?

2.4.3 Ways to manage entropy

My favorite observation from *The Phoenix Project* by Gene Kim, Kevin Behr, and George Spafford[15] is that you only get value from projects when they finish: to make progress, above all else, you must ensure that some of your projects finish.

That might imply that there is an easy solution, but finishing projects is pretty hard when most of your time is consumed by other demands.

Let's tackle hiring first, as hiring and training are often a team's biggest time investment.

When your company has decided that it is going to grow, you cannot stop it from growing, but, on the other hand, you absolutely can concentrate that growth, such that your teams alternate between periods of rapid hiring and periods of consolidation and gelling. Most teams work best when scoped to approximately eight engineers, as each team gets to that point, you can move the hiring spigot to another team (or to a new team). As the post-hiring team gels, eventually the entire group will be trained and able to push projects forward.

You can do something similar on an individual basis, rotating engineers off of interviewing periodically to give them time to recuperate. With high interview loads, you'll sometimes notice last year's solid interviewer giving a poor experience to a candidate or rejecting every

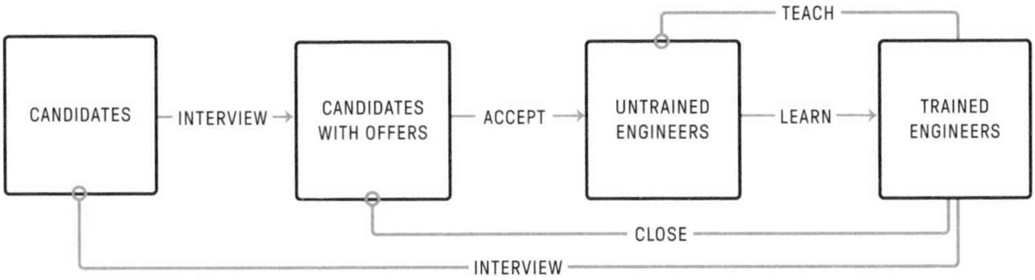

Figure 2.9
Candidates get offers, become untrained, and then learn.

incoming candidate. If your engineer is doing more than three interviews a week, it is a useful act of mercy to give them a month off every three or four months.

I have less evidence on how to tackle the training component of this, but generally you start to see larger companies do major investments in both new-hire bootcamps and recurring education classes.

I'm optimistically confident that we're not entirely cargo-culting this idea from each other, so it probably works, but I hope to get an opportunity to spend more time understanding how effective those programs can be. If you could get training down to four weeks, imagine how quickly you could hire without overwhelming the existing team!

The second most effective time thief that I've found is ad hoc interruptions: getting pinged on HipChat or Slack, taps on the shoulder, alerts from your on-call system, high-volume email lists, and so on.

The strategy here is to funnel interruptions into an increasingly small area, and then automate that area as much as possible. Ask people to file tickets, create chatbots that automate filing tickets, create a service cookbook, and so on.

With that setup in place, create a rotation for people who are available to answer questions, and train your team not to answer other forms of interruptions. This is remarkably uncomfortable because we want to be helpful humans, but it becomes necessary as the number of interruptions climbs higher.

One specific tool that I've found extremely helpful here is an ownership registry, which allows you to look up who owns what, eliminating the frequent "Who owns X?" variety of question. You'll need this sort of thing to automate paging the right on-call rotation, so you might as well get two useful tools out of it!

A similar variant of this is ad hoc meeting requests. The best tool that I've found for this is to block out a few large chunks of time each week to focus. This can range from telecommuting on Thursday, to blocking out Monday and Wednesday afternoons, to blocking out from 8–11 each morning. Experiment a bit and find something that works well for you.

Finally, the one thing that I've found at companies with very few interruptions and have observed almost nowhere else: really great, consistently available documentation. It's probably even harder to bootstrap documentation into a non-documenting company than it is to bootstrap unit tests into a non-testing company, but the best solution to frequent interruptions I've seen is a culture of documentation, documentation reading, and a documentation search that actually works.

There are a non-zero number of companies that do internal documentation well, but I'm less sure if there are a non-zero number of companies with more than 20 engineers that do this well. If you know any, please let me know so that I can pick their brains.

In my opinion, probably the most important opportunity is designing your software to be flexible. I've described this as "fail open and layer policy"; the best system rewrite is the one that didn't happen, and if you can avoid baking in arbitrary policy decisions that will change frequently over time, then you are much more likely to be able to keep using a system for the long term.

If you're going to have to rewrite your systems every few years due to increased scale, let's avoid any unnecessary rewrites, ya know?

Along these lines, if you can keep your interfaces generic, then you are able to skip the migration phase of system re-implementation, which tends to be the longest and trickiest phase, and you can iterate much more quickly and maintain fewer concurrent versions. There is absolutely a cost to maintaining this extra layer of indirection, but if you've already rewritten a system twice, take the time to abstract the interface as part of the third rewrite and thank yourself later. (By the time you'd do the fourth rewrite, you'd be dealing with migrating six times as many engineers.)

Finally, a related antipattern is the gatekeeper pattern. Having humans who perform gatekeeping activities creates very odd social dynamics, and is rarely a great use of a human's time. When at all possible, build systems with sufficient isolation that you can allow most actions to go forward. And when they do occasionally fail, make sure that they fail with a limited blast radius.

There are some cases in which gatekeepers are necessary for legal or compliance reasons, or because a system is deeply frail, but I think that we should generally treat gatekeeping as a significant implementation bug rather than as a stability feature to be emulated.

2.4.4 Closing thoughts

None of the ideas here are instant wins. It's my sense that managing rapid growth is more along the lines of stacking small wins than identifying silver bullets. I have used all of these techniques, and am using most of them today to some extent or another, so hopefully they will at least give you a few ideas. Something that is somewhat ignored a bit here is how to handle urgent project requests when you're already underwater with your existing work and maintenance. The most valuable skill in this situation is learning to say no in a way that is appropriate to your company's culture. That probably deserves its own chapter. There are probably some companies where saying no is culturally impossible, and in those places I guess you either learn to say your noes as yeses, or maybe you find a slightly easier environment to participate in.

How do you remain productive in times of hypergrowth?

2.5 Where to stash your organizational risk?

Lately, I'm increasingly hearing folks reference the idea of *organizational debt*. This is the organizational sibling of *technical debt*, and it represents things like biased interview processes and inequitable compensation mechanisms. These are systemic problems that are preventing your organization from reaching its potential. Like technical debt, these risks linger because they are never the most pressing problem. Until that one fateful moment when they are.

Within organizational debt, there is a volatile subset most likely to come abruptly due, and I call that subset *organizational risk*. Some good examples might be a toxic team culture, a toilsome fire drill, or a struggling leader.

These problems bubble up from your peers, skip-level one-on-ones,[16] and organizational health surveys. If you care and are listening, these are hard to miss. But they are slow to fix. And, oh, do they accumulate! The larger and older your organization is, the more you'll find perched on your capable shoulders.

How you respond to this is, in my opinion, the core challenge of leading a large organization. How do you continue to remain emotionally engaged with the challenges faced by individuals you're responsible to help, when their problem is low in your problems queue? In that moment, do you shrug off the responsibility, either by changing roles or picking powerlessness? Hide in indifference? Become so hard on yourself that you collapse inward?

I've tried all of these! They weren't very satisfying.

What I've found most successful is to identify a few areas to improve, ensure you're making progress on those, and give yourself permission to do the rest poorly. Work with your manager to write this up as an explicit plan and agree on what reasonable progress looks like. These issues are still stored with your other bags of risk and responsibility, but you've agreed on expectations.

Now you have a set of organizational risks that you're pretty confident will get fixed, and then you have all the others: known problems, likely to go sideways, that you don't believe you're able to address quickly. What do you do about those?

I like to keep them close.

Typically, my organizational philosophy is to stabilize team-by-team and organization-by-organization. Ensuring any given area is well on the path to health before moving my focus. I try not to push risks onto teams that are functioning well. You do need to delegate some risks, but generally I think it's best to only delegate solvable risk. If something simply isn't likely to go well, I think it's best to hold the bag yourself. You *may* be the best suited to manage the risk, but you're almost certainly the best positioned to take responsibility.

As an organizational leader, you'll always have a portfolio of risk, and you'll always be doing very badly at some things that are important to you. That's not only okay, it's unavoidable.

2.6 Succession planning

Two or three years into a role, you may find that your personal rate of learning has trailed off. You know your team well, the industry particulars are no longer quite as intimidating, and you have solved the mystery of getting things done at your company. This can be a sign to start looking for your next role, but it's also a great opportunity to build experience with succession planning.

Succession planning is thinking through how the organization would function without you, documenting those gaps, and starting to fill them in. It's awkward enough to talk about that it doesn't get much discussion, but it's a foundational skill for building an enduring organization.

2.6.1 What do you do?

The first step in succession planning is to figure out what you do. This seems like it should be easy, but I've found it surprisingly hard! There are the obvious things you do—one-on-ones, meetings, head count planning—but you're probably filling in a hundred little holes that you don't even think about.

The approach I've taken is to consider your work from several different angles:

- Take a look at your calendar and write down **your role in meetings**. This goes for explicit roles, like owning a meeting's agenda, and also for more nuanced roles, like being the first person to champion others' ideas, or the person who is diplomatic enough to raise difficult concerns.

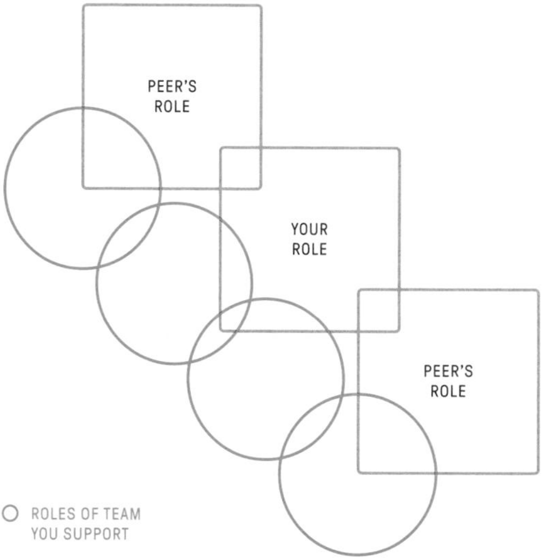

Figure 2.10
Succession planning.

- Take a second pass on your calendar for non-meeting stuff, like interviewing and closing candidates.

- Look back over the past six months for **recurring processes**, like roadmap planning, performance calibrations, or head count decisions, and document your role[17] in each of those processes.

- For each of the **individuals you support**, in which areas are your skills and actions most complementary to theirs? How do you help them? What do they rely on you for? Maybe it's authorization, advice navigating the organization, or experience in the technical domain.

- Audit inbound chats and emails for requests and questions coming your way.

- If you keep a **to-do list**, look at the categories of the work you've completed over the past six months, as well as the stuff you've been wanting to do but keep putting off.

- Think through the **external relationships** that have been important for you in your current role. What kinds of folks have been important, and who are the strategic partners that someone needs to know?

  After exploring each of these avenues, you'll have quite a long list of things. Test the list on a few folks whom you work closely with and see if you've missed anything. Congratulations, now you know what your job is!

2.6.2 Close the gaps

Take your list, and for each item try to identify the individuals who could readily take on that work. Good job, cross those out.

For items without someone who is ready today, identify a handful of individuals who could potentially take it over. (Depending on the size of your list, it may be helpful to cluster similar items into groups to reduce the toil of running this exercise.)

If you're working at a well-established company, you may find that there aren't too many gaps that couldn't be readily filled by someone else. However, if you're at a company going through hypergrowth,[18] it's common to find that everyone is already working in the most complex role of their career, and you'll uncover gaps, gaping and cavernous.

Filter the gaps down to two lists:

1. The first should cover the *easiest gaps to close*. Maybe it'll require a written document or a quick introduction. You should be able to close one of these in less than four hours.

2. The latter will be the *riskiest gaps*. These are the areas where you're uniquely valuable to the company, where other folks are missing skills, and where getting the tasks done is truly important. You'd expect closing one of these gaps to require ongoing effort over several months.

Write up a plan to close *all* of the easy gaps and *one or two* of the riskiest gaps. Add it to your personal goals, and then, congrats, you've completed a round of succession planning!

This isn't a one-time tool, but rather a great exercise to run once a year to identify things you could be delegating. This helps nurture an enduring organization, and also frees up time for you to continue growing into a larger role as well. You can even get a sense of how well you're doing by taking a two- or three-week vacation and seeing what slips through the cracks.

Those items can be the start of next year's list!

3

CHAPTER 3

Tools

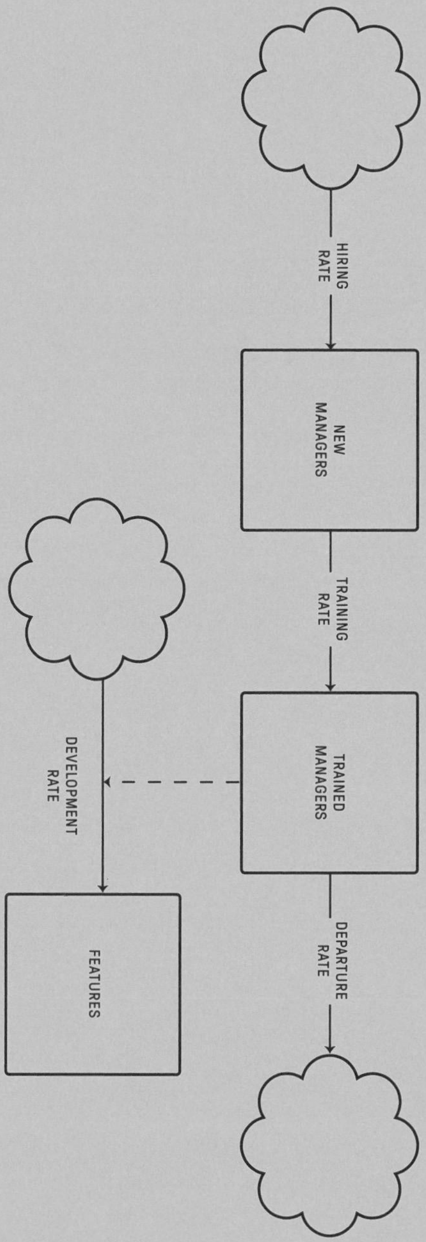

Figure 3.1
System diagram for hiring and training new managers.

Tools

If you ask a manager about their proudest moments, they will probably tell you a story about helping someone grow. If you ask that same manager about their most challenging experience, they will probably talk about a layoff, a reorganization, a shift in company direction, or the time they weathered an economic downturn. In management, change is the catalyst of complexity.

The best changes often go unnoticed, moving from one moment of stability to another, with teams and organizations feeling stable at every step. The key tools for leading efficient change are systems thinking, metrics, and vision. When the steps of change are too wide, teams get destabilized, and gaps open within them. In those moments, managers create stability by becoming glue. We step in as product managers, program managers, recruiters, or salespeople to hold the bits together until an expert relieves us.

This chapter provides a box of tools for managing change, both from the abstract chair of guiding change and from the more visceral role of serving as glue during periods of transition.

3.1 Introduction to systems thinking

Many effective leaders I've worked with have the uncanny knack for working on leveraged[1] problems. In some problem domains, the product management skill set[2] is extraordinarily effective for identifying useful problems, but systems thinking is the most universally useful tool kit I've found.

If you really want a solid grasp on systems thinking fundamentals, you should read *Thinking in Systems: A Primer*[3] by Donella H. Meadows, but I'll do my best to describe some of the basics and to work through a recent scenario in which I found the systems thinking approach to be exceptionally useful.

3.1.1 Stocks and flows

The fundamental observation of systems thinking is that the links between events are often more subtle than they appear. We want to

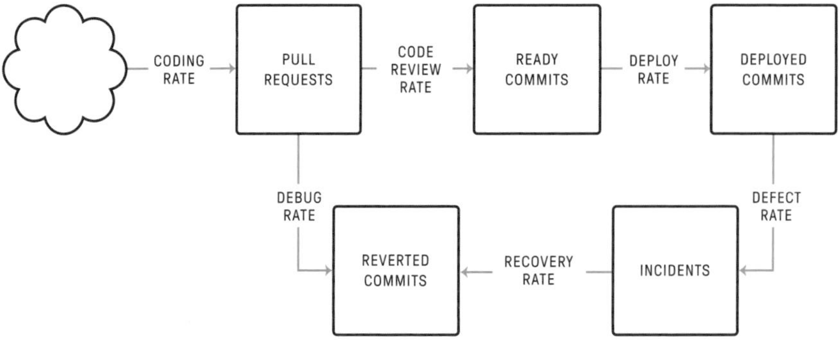

Figure 3.2
System diagram for developer productivity.

describe events causally—our managers are too busy because we're trying to ship our current project—but few events occur in a vacuum.

Big changes appear to happen in a moment, but if you look closely underneath the big change, there is usually a slow accumulation of small changes. In this example, perhaps the managers are busy because no one hired and trained the managers required to support this year's project deadlines. These accumulations are called *stocks*, and are the memory of changes over time. A stock might be the number of trained managers at your company.

Changes to stocks are called *flows*. These can be either *inflows* or *outflows*. Training a new manager is an inflow, and a trained manager who departs the company is an outflow. Diagrams in this chapter represent flows with solid dark lines.

The other relationship, represented in figure 3.1 by a dashed line, is an *information link*. This indicates that the value of a stock is a factor in the size of a flow. The link here shows that the time available for developing features depends on the number of trained managers.

Often, a stock outside of a diagram's scope will be represented as a cloud, indicating that something complex happened there that we're not currently exploring. It's best practice to label every flow, and to keep in mind that every flow is a rate, whereas every stock is a quantity.

3.1.2 Developer velocity

When I started thinking of an example of the usefulness of systems thinking, one came to mind immediately. Since reading *Accelerate: The Science of Lean Software and DevOp*, by Gene Kim, Jez Humble, and Nicole Forsgren,[4] I've spent a lot of time pondering the authors' definition of velocity.

They focus on four measures of developer velocity:

1. **Delivery lead time** is the time from the creation of code to its use in production.

2. **Deployment frequency** is how often you deploy code.

3. **Change fail rate** is how frequently changes fail.

4. **Time to restore service** is the time spent recovering from defects.

The book uses surveys from tens of thousands of organizations to assess each one's overall productivity and show how that correlates to the organization's performance on those four dimensions.

These dimensions kind of intuitively make sense as measures of productivity, but let's see if we can model them into a system that we can use to reason about developer productivity:

- **Pull requests** are converted into **ready commits** based on our *code review rate*.

- **Ready commits** convert into **deployed commits** at *deploy rate*.

- **Deployed commits** convert into **incidents** at *defect rate*.

- **Incidents** are remediated into **reverted commits** at *recovery rate*.

- **Reverted commits** are debugged into new **pull requests** at *debug rate*.

Linking these pieces together, we see a *feedback loop*, in which the system's downstream behavior impacts its upstream behavior. With a sufficiently high *defect rate* or slow *recovery rate*, you could easily see a world where each deploy leaves you even further behind.

If your model is a good one, opportunities for improvement should be immediately obvious, which I believe is true in this case. However, to truly identify where to invest, you need to identify the true values of these stocks and flows! For example, if you don't have a backlog of **ready commits**, then speeding up your *deploy rate* may not be valuable. Likewise, if your *defect rate* is very low, then reducing your *recovery time* will have little impact on the system.

Creating an arena for quickly testing hypotheses about how things work, without having to do the underlying work beforehand, is the aspect of systems thinking that I appreciate most.

3.1.3 Model away

Once you start thinking about systems, you'll find that it's hard to stop. Pretty much any difficult problem is worth trying to represent as a system, and even without numbers plugged in I find them powerful thinking aids.

If you do want the full experience, there are relatively few tools out there to support you. Stella[5] is the gold standard, but the price is quite steep, with a nonacademic license costing more than a new laptop. The best cheap alternative that I've found is Insight Maker,[6] which has some UI quirks but features a donation-based payment model.

3.2 Product management: exploration, selection, validation

Most engineering organizations separate engineering and product leadership into distinct roles. This is usually ideal, not only because these roles benefit from distinct skills but also because they thrive on different perspectives and priorities. It's quite hard to do both well at the same time.

I've met many product managers who are excellent operators, but few product managers who can operate at a high degree while also getting deep into their users' needs. Likewise, I've worked with many engineering managers who ground their work in their users' needs, but I've known few who can fix their attention on those users when things start getting rocky within their team.

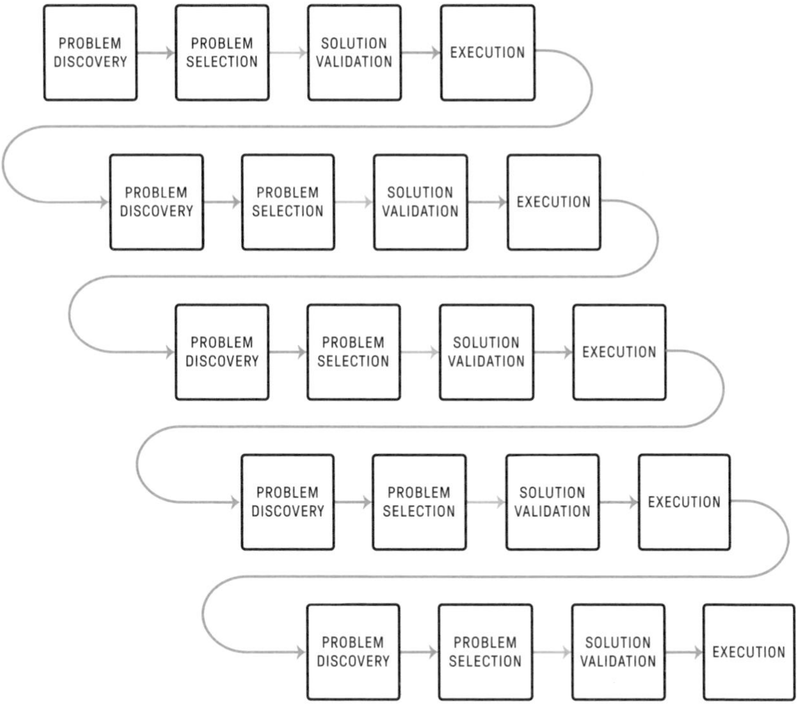

Figure 3.3
Iterative process of product development.

Reality isn't always accommodating of this ideal setup. Maybe your team's product manager leaves or a new team is being formed,[7] and you, as an engineering leader, need to cover both roles for a few months. This can be exciting, and, yes this can be a time when "exciting" rhymes with "terrifying."

Product management is a deep profession, and mastery requires years of practice, but I've developed a simple framework to use when I've found myself fulfilling product management[8] responsibilities for a team. It's not perfect, but hopefully it'll be useful for you as well.

Product management is an iterative elimination tournament, with each round consisting of *problem discovery*, *problem selection*, and *solution validation*. *Problem discovery* is uncovering possible problems to work on, *problem selection* is filtering those problems down to a viable subset, and *solution validation* is ensuring that your approach to solving those problems works as cheaply as possible.

If you do a good job at all three phases, you win the luxury of doing it all again, this time with more complexity and scope. If you don't do well, you end up forfeiting or being asked to leave the game.[9]

3.2.1 Problem discovery

The first phase of a planning cycle is exploring the different problems that you could pick to solve. It's surprisingly common to skip this phase, but that, unsurprisingly, leads to inertia-driven local optimization. Taking the time to evaluate which problem to solve is one of the best predictors I've found of a team's long-term performance.

The themes that I've found useful for populating the problem space are:

Users' pain. What are the problems that your users experience? It's useful to go broad via survey mechanisms, as well as to go deep by interviewing a smaller set of interesting individuals across different user segments.

Users' purpose. What motivates your users to engage with your systems? How can you better enable users to accomplish their goals?

Benchmark. Look at how your company compares to competitors in the same and similar industries. Are there areas in which you are quite weak? Those are areas to *consider* investing in. Sometimes folks keep to a narrow lens when benchmarking, but I've found that you learn the most interesting things by considering both fairly similar and rather different companies.

Cohorts. What is hiding behind your clean distributions? Exploring your data for the cohorts hidden behind top-level analysis is an effective way to discover new kinds of users with surprising needs.

Competitive advantages. By understanding the areas you're exceptionally strong in, you can identify opportunities that you're better positioned to fill than other companies.

Competitive moats. Moats are a more extreme version of a competitive advantage. Moats represent a sustaining competitive advantage, which makes it possible for you to pursue offerings that others simply cannot. It's useful to consider moats in three different ways:

- What do your existing moats enable you to do today?
- What are the potential moats you could build for the future?
- What moats are your competitors luxuriating behind?

Compounding leverage. What are the composable blocks you could start building today that would compound into major product or technical leverage[10] over time? I think of this category of work as finding ways to get the benefit at least twice. These are potentially tasks that initially don't seem important enough to prioritize, but whose compounding value makes the work possible to prioritize.

- A design example might be introducing to an application a new navigation scheme that better supports the expanded set of actions and modes you have today, and that will support future proliferation as well. (Bonus points if it manages to prevent future arguments about the positioning of new actions relative to existing ones!)
- An infrastructure example might be moving a failing piece of technology to a new standard. This addresses a reliability issue, lowers maintenance costs, and also reduces the costs of future migrations.[11]

3.2.2 Problem selection

Once you've identified enough possible problems, the next challenge is to narrow down to a specific problem portfolio. Some of the aspects that I've found useful to consider during this phase are:

Surviving the round. Thinking back to the iterative elimination tournament, what do you need to do to survive the current round? This might be the revenue that the product will need to generate to avoid getting canceled, adoption, etc.

Surviving the next round. Where do you need to be when the next round in order to avoid getting eliminated then? There are a number of ways (many of them revolving around quality trade-offs) to reduce long-term throughput in favor of short-term velocity. (Conversely, winning leads to significantly more resources later, so that trade-off is appropriate sometimes!)

Winning rounds. It's important to survive every round, but it's also important to eventually win a round! What work would ensure that you're trending toward winning a round?

Consider different time frames. When folks disagree about which problems to work on, I find that the conflict is most frequently rooted in different assumptions about the correct time frame to optimize for. What would you do if your company was going to run out of money in six months? What if there were no external factors forcing you to show results until two years out? Five years out?

Industry trends. Where do you think the industry is moving, and what work will position you to take advantage of those trends, or to at least avoid having to redo the work in the near future?

Return on investment. Personally, I think people often under-prioritize quick, easy wins. If you're in the uncommon position of understanding both the impact and costs of doing small projects, then take time to try ordering problems by expected return on investment. At this phase, you're unlikely to know the exact solution, so figuring out cost is tricky, but for categories of problems that you've seen before you can probably make a solid guess. (If you don't personally have relevant experience, ask around.) Particularly in cases where wins are compounding, they may be surprisingly valuable over the medium and long term.

Experiments to learn. What could you learn now that would make problem selection in the future much easier?

3.2.3 Solution validation

Once you've narrowed down the problem you want to solve, it's easy to jump directly into execution, but that can make it easy to fall in love with a difficult approach. Instead, I've found that it's well worth it to take the risk out of your approach with an explicit solution validation phase.

The elements that I've found effective for solution validation are:

Write a customer letter. Write the launch announcement that you would send after finishing the solution. Are you able to write something exciting, useful, and real? It's much more useful to test it against your actual users than to rely on your intuition.

Identify prior art. How do peers across the industry approach this problem? The fact that others have solved a problem in a certain way doesn't mean that it's a great way, but it does at least mean it's possible. A mild caveat: it's better to rely on people you have some connection to instead of on conference talks and such, since there is a surprisingly large amount of misinformation out there.

Find reference users. Can you find users who are willing to be the first users for the solution? If you can't, you should be skeptical whether what you're building is worthwhile.

Prefer experimentation over analysis. It's far more reliable to get good at cheap validation than it is to get great at consistently picking the right solution. Even if you're brilliant, you are almost always missing essential information when you begin designing. Analysis can often uncover missing information, but it depends on knowing where to look, whereas experimentation allows you to find problems that you didn't anticipate.

Find the path more quickly traveled. The most expensive way to validate a solution is to build it in its entirety. The upside of that approach is that you've lost no time if you picked a good solution. The downside is that you've sacrificed a huge amount of time if it's not. Try to find the cheapest way to validate.

Justify switching costs. What will the costs of switching be for users who move to your solution? Even if folks want to use it, high switching costs may mean that they simply won't be able to. Test with your potential users if they'd be willing to pay the full cost of migrating to your solution instead of their existing planned work.

As an aside, I've found that most aspects of running a successful technology migration[12] overlap with good solution validation! This is a very general skill that will repay many times over the time you invest in learning it.

Putting these three elements in place today—exploration, selection, and validation—won't make you an exceptional product manager overnight, but they will provide a solid starting place to develop those skills and perspective for the next time you find yourself donning the product manager hat.

3.3 Visions and strategies

As an organization grows beyond 50 people or so, you'll feel a building pressure to add a third layer of management, and eventually you will. This ought to be a benign event: What's the difference between supporting some managers and supporting their managers? It shouldn't be too different, but for me it was when my previous mechanisms of alignment stopped working very well.

Where I was once partnering with teams on their project roadmaps, now I found myself increasingly surprised by the projects that they were working on. Where I was once debating different approaches with the teams, now that conversation was happening in rooms I didn't have time to join.

My first instinct was to dive in and understand each instance, but that, unsurprisingly, wasn't a very scalable solution. My second instinct was to design a series of "operating reviews" in which we periodically reviewed metrics and major projects. While those were useful in their own right, they proved more effective for learning and fine-tuning than for broad alignment. Being out of alignment for a quarter is just so much uncaptured potential.

What I needed was a way to coordinate my approach across teams, both in terms of very specific challenges and in terms of our long-term direction. After experimenting with a handful of different approaches, agreeing on strategy and vision has been the most effective approach that I've found to alignment at scale.

3.3.1 Strategies and visions

Strategies are grounded documents which explain the trade-offs and actions that will be taken to address a specific challenge. *Visions* are aspirational documents that enable individuals who don't work closely together to make decisions that fit together cleanly.

Picking the right format for your needs is important, but the most important thing is probably to give both a try and get a feel for them! These are peculiar genres of literature that take some practice to master. It's taken me years to get comfortable writing visions, and it was only when I started to support teams with seemingly incompatible ideologies that these documents value became clear. The same

	STRATEGY	VISION
PURPOSE	APPROACH TO A SPECIFIC CHALLENGE	A GENTLE, ALIGNING PRESSURE
CHARACTER	PRACTICAL	ASPIRATIONAL
TIME FRAME	VARIABLE	LONG-TERM
SPECIFICITY	ACCURATE, DETAILED	ILLUSTRATIVE, DIRECTIONAL
QUANTITY	AS MANY AS USEFUL	AS FEW AS POSSIBLE

Figure 3.4
Table of differences between strategy and vision.

goes for writing strategies: it took a long time and skeptical study of several strategy books before what I wrote started to come together into a useful artifact.

With that said, it's time to dig into how to write strategies first, and then visions.

3.3.2 Strategy

A *strategy* recommends specific actions that address a given challenge's constraints. A structure that I've found extremely effective[13] is described in *Good Strategy/Bad Strategy* by Richard Rumelt,[14] and has three sections: *diagnosis*, *policies*, and *actions*.

The *diagnosis* is a theory describing the challenge at hand. It calls out the factors and constraints that define the challenge, and at its core is a very thorough problem statement. An example of a simple diagnosis might be "I am too busy to think about long-term goals. I attend 35 hours of meetings each week. I am under pressure to immediately increase team performance. I believe that if I stop doing my current meetings, short-term team performance will decrease. I am concerned that if my short-term team performance decreases, I may lose face as an effective leader, which will undermine my career opportunities. I believe that if I don't think about long-term goals, our performance will never improve, which will also undermine my career opportunities." Before you've even finished reading a great diagnosis, you'll often have identified several good candidate approaches. That's the power of a well-defined problem statement, and why it's an important foundational element for your strategy.

The second step is to identify *policies* that you will apply to address the challenge. These describe the general approach that you'll take, and are often trade-offs between two competing goals. Continuing the above example, you might choose to allow short-term performance to dip in order to invest into long-term performance, combined with a policy of proactive expectation-setting with your stakeholders. Conversely, you might choose to take a hill-climbing approach to long-term performance, with iterative short-term improvement leading to improved long-term performance. Both are valid guiding policies, and both embrace the reality that you have limited resources to invest. When you read bad guiding policies, you think, "so what?" because its found a way to justify entrenching the status quo. When you read good guiding policies, you think, "Ah, that's really going to annoy Anna, Bill, and Claire," because the approach takes a clear stance on competing goals.

When you apply your guiding policies to your diagnosis, you get your *actions*. Folks are often comfortable with hard decisions in the abstract, but struggle to translate policies into the specific steps to implement them. This is typically the easiest part to write, but publishing it and following through with it can be a significant test of your commitment. In the example above, your specific actions might be to stop attending weekly team meetings in order to free up time and move to a monthly metrics review, combined with blocked-out focus hours that you unapologetically shelter. When you read good, coherent actions, you think, "This is going to be uncomfortable, but I think it can work." When you read bad ones, you think, "Ah, we got afraid of the consequences, and we aren't really changing anything."

Because strategies are specific to a given problem, it's okay—and even encouraged—to write quite a few of them. Over the past year, I've worked with people on strategies for how we partner with other teams, how we manage end-to-end API latency, and how we manage infrastructure costs.[15] I've also peered over others' shoulders as they worked on quite a few more ideas. The act of writing a strategy leads folks through a systematic analysis, so, even if we don't share them, writing these documents helps us work through quite a few challenges, both overwhelming and mundane.

People sometimes describe strategy as artful or sophisticated, but I've found that the hardest part of writing a good strategy is pretty mundane. You must be honest about the constraints that are making the challenge difficult, which almost always include people and organizational aspects that are uncomfortable to acknowledge. No extent of artistry can solve a problem that you're unwilling to admit.

3.3.3 Vision

If strategies describe the harsh trade-offs necessary to overcome a particular challenge, then *visions* describe a future in which those trade-offs are no longer mutually exclusive. An effective vision helps folks think beyond the constraints of their local maxima, and lightly aligns progress without requiring tight centralized coordination.

You should be writing from a place far enough out that the error bars of uncertainty are indisputably broad, where you can focus on the concepts and not the particulars. Visions should be detailed, but the details are used to illustrate the dream vividly, not to prescriptively constrain its possibilities.

A good vision is composed of:

1. **Vision statement:** A one- or two-sentence aspirational statement to summarize the rest of the document. This is your core speaking point, which you will repeat at each meeting, planning period, and strategy review. It shouldn't try to capture every detail of your vision, but it does need to memorably evoke your vision effectively.

2. **Value proposition:** How will you be valuable to your users and to your company? What kinds of success will you enable them to achieve? There is a sequencing question of whether you should start with *capabilities* (the next bullet) and reason into *value proposition* or do the opposite, but I've found that starting from your users leads to visions that are both more ambitious and more grounded.

3. **Capabilities:** What capabilities will the product, platform, or team need in order to deliver on your value proposition? Will it need to support multiple independent business lines? Will it need to deliver against the disparate needs of several distinct customer cohorts?

4. **Solved constraints:** What are the constraints that you're limited by today, but that in the future you'll no longer be constrained by? For example, if you're currently "spending into" developer velocity, perhaps in the future you'll be able to achieve significant developer velocity while also maintaining low costs.

5. **Future constraints:** What are the constraints that you expect to encounter in this wonderful future? Hopefully, these new constraints will be things that are easy to adjust, like funding or hiring.

6. **Reference materials:** Link all the existing plans, metrics, updates, references, and documents into an appendix for those who want to understand more of the thinking that went into the vision. This allows you to shed complexity from your vision document without sacrificing context.

7. **Narrative:** Once you've written the previous sections, the last step of writing a compelling vision is to synthesize all those details into a short—maybe one-page—narrative that serves as an easy-to-digest summary. In your final document, this is probably the second section, following the statement.

Put all these pieces together, and you've crafted a document that is a guiding hand to align decisions yet still creates room for teams to make their own choices and trade-offs along the way. You'll know a vision is succeeding when people reference the document to make their own decisions, and you'll know it's struggling when decisions keep happening that don't fit into its direction.

Compared to strategies, there is more room to play with the vision format. You'll be reading far fewer visions than strategies, so consistency is less important. Play with the content a bit to find what works best for you.

A few additional tips that I've found especially useful:

Test the document! This is a core leadership tool, and your first version will almost certainly be bad. Write a draft, sit down with a few different folks to get their perspectives, then iterate. Keep doing this until you've synthesized feedback. If there is feedback you disagree with, embrace the vision as an opportunity to address conflict by explicitly acknowledging disagreements within the vision text.

Refresh periodically. Take some time every year to refresh the vision, and prefer usefulness over consistency. If your old vision doesn't resonate, it's okay to start over: it's a sign that you've learned a lot over the past year.

Use present tense. This makes the writing impactful and concise, and conveys a sense of confidence about the future.

Write simply. Often, visions are saturated with buzzwords, which turns readers off.

You'll likely want *one vision for every complete distinct area, but no more.* If areas overlap, you get the alignment value from having one unified vision; having two clearly articulated visions in one place is worse than having zero.

Like other leadership tools, a vision or strategy document is a solution to a specific set of problems, and it's not always useful. If your team is aligned and doing good work, time spent writing these probably won't be too valuable.

However, if your team is struggling to align with stakeholders, or if you're struggling to lead a cohesive organization, these documents are exceptionally useful, fairly quick to write as you gain practice, and low risk (at worst, they get ignored).

3.4 Metrics and baselines

There is a moment in every company's growth when top-level planning shifts from discussing specific projects to talking about goals. This happens recursively across each scope of leadership, as areas of accountability become too broad or complex for their leaders to consistently understand every project's details.

This can be a very empowering moment because goals decouple the "what" from the "how," but it can also be a confusing transition for everyone involved: writing clear goals takes a bit of practice.

⊙ Defining goals

Bad goals are indistinguishable from numbers. "Our p50 build time will be below two seconds," or "We'll finish eight large projects." You'll know a goal is just a number when you read it and aren't sure if it's ambitious or whether it matters.

Good goals are a composition of four specific kinds of numbers:

1. A **target** states where you want to reach.
2. A **baseline** identifies where you are today.
3. A **trend** describes the current velocity.
4. A **time frame** sets bounds for the change.

Put these all together, and a well-structured goal takes the form of: "In Q3, we will reduce time to render our frontpage from 600ms (p95) to 300ms (p95). In Q2, render time increased from 500ms to 600ms."

The two tests of an effective goal are whether someone who doesn't know much about an area can get a feel for a goal's degree of difficulty, and whether afterward they can evaluate if it was successfully achieved. If you define all four aspects, typically your goal will fulfill both criteria.

⊙ Investments and baselines

There are two particularly interesting kinds of goals: investments and baselines. Investments describe a future state that you want to reach, and baselines describe aspects of the present that you want to preserve.

Imagine that you wanted to speed up your data pipeline. Your goal might be, "Core batch jobs should finish within three hours (p95) by the end of Q3. They currently take six hours (p95), and over the course of Q2 they got two hours slower." This is a well-structured goal, but it's also incomplete because you could likely reach that goal tomorrow by doubling the size of your cluster, which is probably not a desirable outcome.

The best way to avoid such unintended outcomes is to pair your investment goals with baseline metrics, sometimes referred to as countervailing metrics. For the data pipeline example, a few of the baseline metrics might be:

- Efficiency of running core batch jobs should not exceed current price of $0.05 per GB.
- Core batch jobs should not increase alert load on teams operating or using the pipeline, which are currently alerting twice per week.

Baseline metrics are useful for narrowing the solution space that you explore in order to accomplish your investment goals. They are also useful for identifying when you should pause pursuing your goals and instead invest in platform quality. For example, if you were making excellent progress toward launching a new feature but site stability has regressed below your baselines, this framework provides a structure to trigger rebalancing your priorities.

Although your baselines will often be about preserving a current property, you can also decide to accept some degradation before you want to trigger reprioritization. Perhaps you're okay with costs increasing by 10 percent as long as your investment goals are accomplished. This kind of upfront clarity around trade-offs can be quite powerful.

⊙ Plans and contracts

The most common way to use goals is during a planning process. By agreeing on the mix of investment and baseline goals for each team, you're able to set clear expectations for a team while still giving them full ownership of how they'll satisfy the constraints. I've found that you should specify as few investment goals as possible, maybe three, and that those should be the focus of planning discussions.

You'll probably want to identify more baseline goals than investment goals, but it's easiest to separate them out to avoid bogging down the conversation. Ideally, baselines are carried over across planning periods, such that they frame the investment goals but don't require too much active discussion during any given planning cycle.

One potential exception is when you're using a baseline as a contract with a second party, possibly specifying an SLO,[16] at which point you'll probably want to discuss it more explicitly than other baselines:

missing an SLO will probably require immediate reprioritization, whereas missing most other baselines can generally be addressed more methodically.

From OKRs[17] onward, there are dozens of different approaches to setting metrics, but I've found this format to be a useful, lightweight structure to start from. If folks have found other approaches easier or more useful, I'd love to hear from you!

3.5 Guiding broad organizational change with metrics

Although people often talk about goals and metrics[18] when they're writing new plans or reflecting on past plans, my fondest memories of metrics are when I've seen them used to drive large organizational change. In particular, I've found metrics to be an extremely effective way to lead change with little or no organizational authority, and I wanted to write up how I've seen that work.

At both Stripe and Uber, I've had the opportunity to manage infrastructure costs. (Let me insert a plug for Ryan Lopopolo's amazing blog post on "Effectively Using AWS Reserved Instances."[19]) Folks who haven't thought about this problem often default to viewing it as boring, but I've found that as you dig into it, it's rich soil for learning about leading organizational change.

It's also a good example of how to lead change with metrics!

Infrastructure cost is a great example of a baseline metric.[20] When you're asked to take responsibility for a company's overall infrastructure costs, you're going to start from a goal along the lines of "Maintain infrastructure costs at their current percentage of net revenue of 30 percent." (That percentage is a fictional number for this example's purposes, since the percentage will depend on your industry and maturity, but I *have* found that tying it against net revenue is more useful than pinning it at a specific dollar amount.)

From there, the approach that I've found effective is:

1. **Explore:** The first step is to get data in an explorable format in your data warehouse, an SQL database, or even an Excel spreadsheet. Once there, spend time looking through it and getting a feel for it.

Your goal in this phase is to identify where the levers for change are. For example, you might find that your batch pipeline is the majority of your costs and that your data warehouse is surprisingly cheap, which will allow you to focus further efforts.

2. **Dive:** Once you know the three or four major contributors, go deep on understanding those areas and the levers that drive them. Batch costs might be sensitive to number of jobs, total data stored, or new product development, *or* it might be entirely driven by a couple of expensive jobs. Diving deep helps you build a mental model, and it also kicks off a relationship between you and the teams who you'll want to partner with most closely.

3. **Attribute:** For most company-level metrics (cost, latency, development velocity, etc.), the first step of diving will uncover one team who are nominally accountable for the metric's performance, but they are typically a cloak. When you pull that cloak aside, that team's performance is actually driven by dozens of other teams. For example, you might have a Cloud engineering team who are accountable for provisioning VMs, but they're not the folks writing the code that runs on those VMs. It's easy to simply pass the cost metric on to that Cloud team, but that's just abdicating responsibility to them. What's more useful is to help them build a system of second-degree attribution, allowing you to build data around the teams using the platform. This second degree of attribution is going to allow you to target the folks who can make an impact in the next step.

4. **Contextualize:** Armed with the attribution data, start to build context around each team's performance. The most general and self-managing tool for this is benchmarking. It's one thing for a team to know that they're spending $100,000 a month, and it's something entirely different for them to know that they're spending $100,000 a month *and* that their team spends the second-highest amount out of 47 teams. Benchmarking is particularly powerful because it automatically adapts to changes in behavior. In some cases, benchmarking against all teams might be too coarse, and it may be useful to benchmark against a small handful of cohorts. For example, you might want to define cohorts for front-end, back-end, and infrastructure teams, given that they'll have very different cost profiles.

5. **Nudge:** Once you've built context around the data so that folks can interpret it, the next step is to start nudging them to action! Dashboards

are very powerful for analysis, but the challenge for baseline metrics is that folks shouldn't need to think about them the vast majority of the time, and that can lead to them forgetting about the baselines entirely. What I've found effective is to send push notifications, typically email, to teams whose metric has changed recently, both in terms of absolute change and in terms of their benchmarked performance against their cohort. This ensures that each time you push information to a team, it includes important information that they should act on! What's so powerful about nudges is that simply letting folks know their behavior has changed will typically stir them to action, and it doesn't require any sort of organizational authority to do so. (For more on this topic, take a look at *Nudge* by Richard H. Thaler and Cass R. Sunstein.)[21]

6. **Baseline:** In the best case, you'll be able to drive the organizational impact you need with contextualized nudges, but in some cases that isn't quite enough. The next step is to work with the key teams to agree on baseline metrics for their performance. This is useful because it ensures that the baselines are top-of-mind, and it also gives them a powerful tool for negotiating priorities with their stakeholders. In some cases, this does require some organizational authority, but I've found that folks universally want to be responsible. As long as you can find time to sit down with the key teams and explain why the goal is important, it typically doesn't require much organizational authority.

7. **Review:** The final phase, which hopefully you won't need to reach, is running a monthly or quarterly review that looks at each team's performance, and reaching out to teams to advocate for prioritization if they aren't sustaining their agreed-upon baselines. This typically requires an executive sponsor, because teams who aren't hitting their baselines are almost always being prioritized against other goals, and and they your help explaining to their stakeholders why the change is important.

I've seen this approach work and, more importantly, I've found it to be very scalable. It enables a company to concurrently maintain many baseline metrics without overloading its teams. This is largely because this approach focuses on driving targeted change within the key drivers, only requiring involvement from a small subset of teams for any given metric. The approach is also effective because it tries to minimize top-down orchestration in favor of providing information to encourage teams themselves to adjust priorities.

3.6 Migrations: the sole scalable fix to tech debt

The most interesting migration I ever participated in was Uber's migration from Puppet-managed services to a fully self-service provisioning model in which any engineer at the company could spin up a new service in two clicks. Not only could they, they did, provisioning multiple services each day by the time the service was complete, and every newly hired engineer could spin up a service from scratch on their first day.

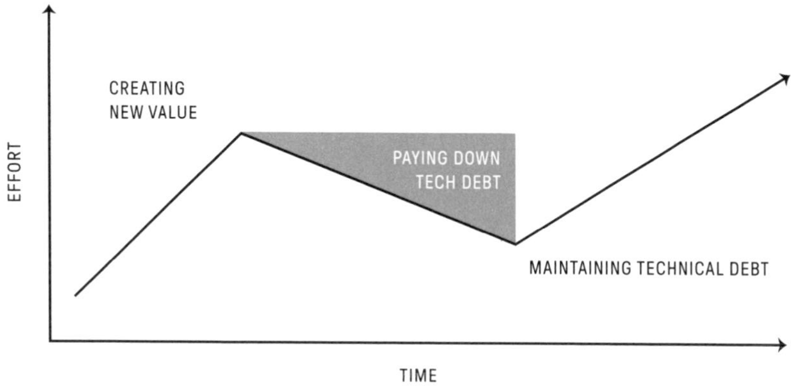

Figure 3.5
Stages of a technical migration.

What made this migration so interesting was the volume. When we started, provisioning a new service took about two weeks of clock time and about two days of engineering time, and we were falling further behind each day. At the time, this was more than just a little stressful, but it was also a perfect laboratory to learn how to run large-scale software migrations: the transistion was large enough to see even small shifts, and long enough that we got to experiment with a number of approaches.

Migrations are both essential and frustratingly frequent as your codebase ages and your business grows: most tools and processes only support about one order of magnitude of growth[22] before becoming ineffective, so rapid growth makes migrations a way of life. This isn't because you have bad processes or poor tools—quite the opposite. The fact that something stops working at significantly increased scale is a sign that it was designed appropriately to the previous constraints rather than being over-designed.[23]

As a result, you switch tools a lot, and your ability to migrate to new software can easily become the defining constraint for your overall velocity. Given their importance, we don't talk about running migrations very often; let's remedy that!

3.6.1 Why migrations matter

Migrations matter because they are usually the only available avenue to make meaningful progress on technical debt.

Engineers hate technical debt. If there is an easy project that they can personally do to reduce tech debt, they'll take it on themselves. Engineering managers hate technical debt, too. If there is an easy project that their team can execute in isolation, they'll get it scheduled. In aggregate, this leads to a dynamic in which there is very little low-hanging fruit to reduce technical debt, and most remaining options require many teams working together to implement them. The result: migrations.

Each migration aims to create technical leverage (Your indexes no longer have to fit on a single server!) or reduce technical debt (Your acknowledged writes are guaranteed to persist a master failover!). They occupy the awkward territory of reduced immediate contribution today in exchange for more capacity tomorrow. This makes migrations controversial to schedule, and as your systems become larger, they become more expensive. Lore tells us that Googlers have a phrase, "running to stand still," to describe a team whose entire capacity is consumed in upgrading dependencies and patterns, such that the group can't make forward progress on the product/system they own. Spending *all* your time on migrations is extreme, but every midsize company has a long queue of migrations that it can't staff: moving from VMs to containers, rolling out circuit-breaking, moving to the new build tool . . . the list extends effortlessly into the sunset.

Migrations are the only mechanism to effectively manage technical debt as your company and code grow. If you don't get effective at software and system migrations, you'll end up languishing in technical debt. (And you'll still have to do one later anyway, it's just that it'll probably be a full rewrite.)

3.6.2 Running good migrations

The good news is that while migrations are hard, there is a pretty standard playbook that works remarkably well: de-risk, enable, then finish.

⊙ De-risk

The first phase of a migration is **de-risking** it, and to do so as quickly and cheaply as possible. Write a **design document** and shop it with the teams that you believe will have the hardest time migrating. Iterate. Shop it with teams who have atypical patterns and edge cases. Iterate. Test it against the next six to twelve months of roadmap. Iterate.

After you've evolved the design, the next step is to **embed into the most challenging one or two teams**, and work side by side with those teams to build, evolve, and migrate to the new system. Don't start with the easiest migrations, which can lead to a false sense of security.

Effective de-risking is essential, because **each team who endorses a migration is making a bet on you** that you're going to get this damn thing done, and not leave them with a migration to an abandoned system that they have to revert to. If you leave one migration partially finished, people will be exceedingly suspicious of participating in the next.

⊙ Enable

Once you've validated the solution that solves the intended problem, it's time to start sharpening your tools. Many folks start migrations by generating tracking tickets for teams to implement, but it's better to slow down and build tooling to programmatically migrate the easy 90 percent.[24] This radically reduces the migration's cost to the broader organization, which increases the organization's success rate and creates more future opportunities to migrate.

Once you've handled as much of the migration programmatically as possible, figure out the **self-service tooling and documentation** that you can provide to allow teams to make the necessary changes without getting stuck. The best migration tools are incremental and reversible: folks should be able to immediately return to previous behavior if something goes wrong, and they should have the necessary expressiveness to de-risk their particular migration path.

Documentation and self-service tooling are products, and they thrive under the same regime: sit down with some teams and watch them follow your instructions, then improve them. Find another team. Repeat. Spending an extra two days intentionally making your documentation clean and your tools intuitive can save years in large migrations. Do it!

⊙ Finish

The last phase of a migration is deprecating the legacy system that you've replaced. This requires getting to 100 percent adoption, and that can be quite challenging.

Start by **stopping the bleeding**, which is ensuring that all newly written code uses the new approach. That can be installing a ratchet in your linters,[25] or updating your documentation and self-service tooling. This is always the first step, because it turns time into your friend. Instead of falling behind by default, you're now making progress by default.

Okay, now you should start **generating tracking tickets**, and set in place a mechanism which **pushes migration status** to teams that need to migrate and to the general management structure. It's important to give wider management context around migrations because the managers are the people who need to prioritize the migrations: if a team isn't working on a migration, it's typically because their leadership has not prioritized it.

At this point, you're pretty close to complete, but you have the long tail of weird or unstaffed work. Your tool now is: **finish it yourself**. It's not necessarily fun, but getting to 100 percent is going to require the team leading the migration to dig into the nooks and crannies themselves.

My final tip for finishing migrations centers around recognition. It's important to celebrate migrations while they're ongoing, but the majority of the celebration and **recognition should be reserved for their successful completion**. In particular, starting but not finishing migrations often incurs significant technical debt, so your incentives and recognition structure should be careful to avoid perverse incentives.

3.7 Running an engineering reorg

I believe that, at quickly growing companies, there are two managerial skills that have a disproportionate impact on your organization's success: making technical migrations cheap, and running clean reorganizations. Do both well, and you can skip that lovely running-to-stand-still sensation, and invest your attention more fruitfully.

Of the two, managing organizational change is more general, so let's work through a lightly structured framework for (re)designing an engineering organization.

Caveat: this is more of a thinking tool than a recipe!

My approach for planning organization change:

1. Validate that organizational change is the right tool.

2. Project head count a year out.

3. Set target ratio of management to individual contributors.

4. Identify logical teams and groups of teams.

5. Plan staffing for the teams and groups.

6. Commit to moving forward.

7. Roll out the change.

Now, let's drill into each of those a bit.

3.7.1 Is a reorg the right tool?

There are two best kinds of reoganizations:

- The one that solves a structural problem.

- The one that you don't do.

There is only one worst kind of reorg: the one you do because you're avoiding a people management issue.

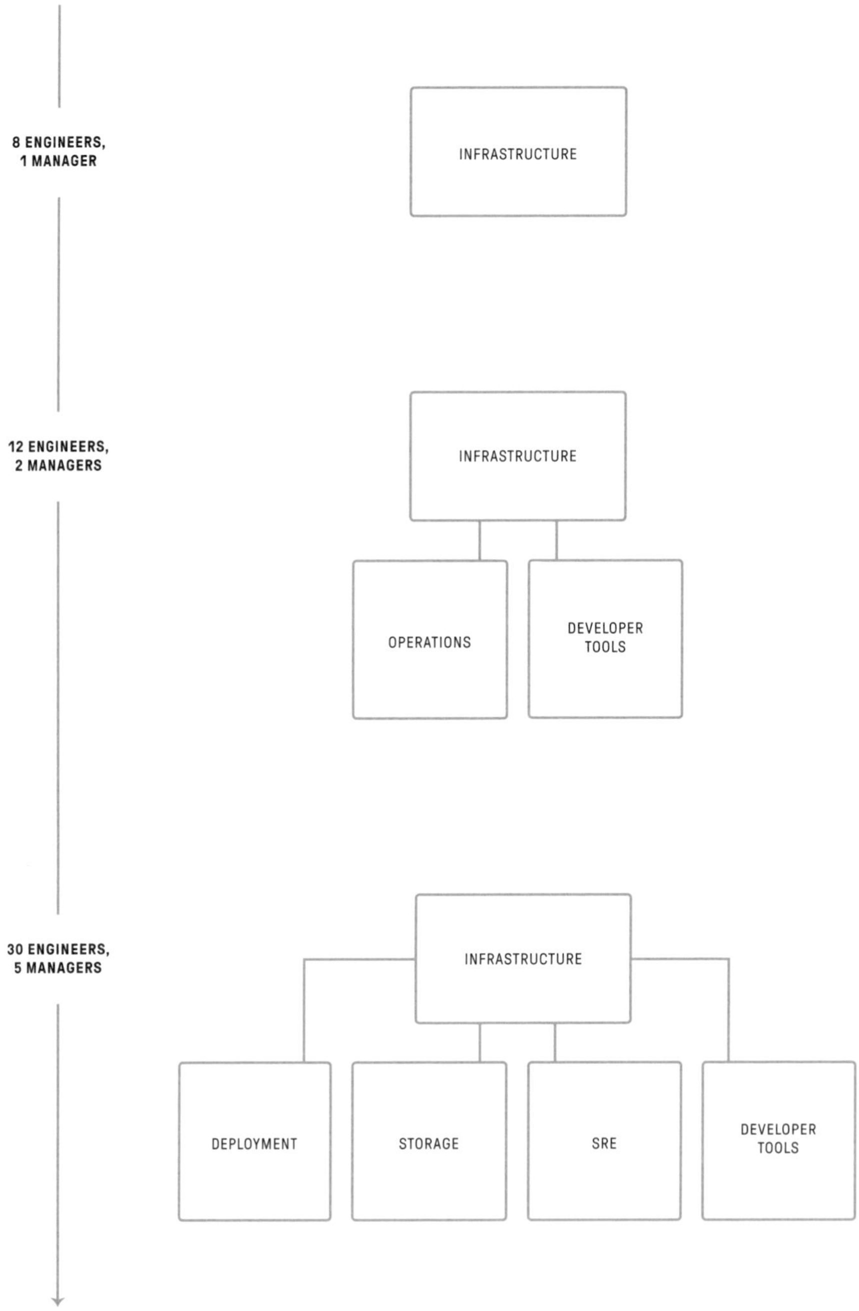

Figure 3.6
Refactoring organizations as they grow.

My checklist for ensuring that a reorganization is appropriate:

1. Is the problem structural? Organization change offers the opportunity to increase communication, reduce decision friction, and focus attention; if you're looking for a different change, consider if there's a more direct approach.

2. Are you reorganizing to work around a broken relationship? Management is a profession where karma *always* comes due, and you'll be better off addressing the underlying issue than continuing to work around it.

3. Does the problem already exist? It's better to wait until a problem actively exists before solving it, because it's remarkably hard to predict future problems. Even if you're right that the problem will occur, you may end up hitting a different problem first.

4. Are the conditions temporary? Are you in a major crunch period or otherwise doing something you don't anticipate doing again? If so, then it may be easier to patch through and rethink on the other side, and avoid optimizing for a transient failure mode.

All right, so you're still thinking that you want a reorg.

3.7.2 Project head count a year out

The first step of designing the organization is determining its approximate total size. I recommend reasoning through this number from three or four different directions:

1. An optimistic number based on what's barely possible.

2. A number based on the "natural size" of your organization, if every team and role was fully staffed.

3. A realistic number based on historical hiring rates.

Then merge those into a single number.

Unless you've changed something meaningful in your process, it's likely that the historical trend will hold accurate, and you should weight that figure the most heavily (and my sense is that the list of easy changes that significantly after hiring outcomes is short).

One of the goals of using the year-out head count number is to avoid optimizing too heavily for your exact current situation and the current individuals you're working with. Organizational change is *so disruptive* to *so many people* that I've increasingly come to believe you should drive organizational design from the boxes and not from the key individuals.

3.7.3 Manager-to-engineer ratio

Once you have your head count projection, you need to identify how many individuals you want each manager to support. This number particularly depends on your company's working definition of an engineering manager's role. If engineering managers are expected to do hands-on technical work, then their teams should likely be three to five engineers (unless the team has been working together well for a long time, in which case things get very specific and hard to generalize about).

Otherwise, targeting five to eight engineers, depending on experience level, is pretty typical. If you're targeting more than eight engineers per manager, then it's worth reflecting on why you believe your managers can support a significantly higher load than industry average: Are they exceptionally experienced? Are your expectations lower than typical?

In any case, pick your target, probably in the six-to-eight range.

3.7.4 Defining teams and groups

Now that you have your target organization size and target ratio of managers to engineers, it's time to figure out the general shape of your organization!

Suppose that you have 35 engineers and 7 engineers per manager.

35 / 7 = 5 managers

$\text{Log}^7(35) \approx 1.8$ managers of managers, or second-degree managers

In a growing company, you should generally round up the number of managers, as this is a calculation "at rest," and your organization will be a living, evolving thing.

Once you have the numbers, these are useful to *ground* you in the general number of teams and groups of teams you should have.

In the first case, with 35 engineers, you're going to want between one and three groups, containing a total of five or six teams. In the latter, with 74 engineers, you'll want two to four groups, comprised of 12 to 15 teams.

Once you've grounded yourself, here are some additional considerations:

1. Can you write a crisp mission statement for each team?

2. Would you personally be excited to be a member of each of the teams, as well as to be the manager of each of those teams?

3. Put teams that work together (especially poorly) as close together as possible. This minimizes the distance for escalations during disagreements, allowing arbiters to have sufficient context. Also, most poor working relationships are the by-product of information gaps, and nothing fills those faster than proximity.

4. Can you define clear interfaces for each team?

5. Can you list the areas of ownership for each team?

6. Have you created a gap-less map of ownership, such that each responsibility is owned by a team? Try to avoid implicitly creating holes of ownership. If you need to create explicit holes of ownership, that's a better solution (essentially, defining unstaffed teams).

7. Are there compelling candidate pitches for each of those teams?

8. As always, are you over-optimizing on individuals versus establishing a sensible structure?

This is the least formulaic aspect of organizational design, and, if possible it's a good time to lean on your network and similar organizations for ideas.

3.7.5 Staffing the teams and groups

With your organization design and head count planning, you can roughly determine when you'll need to fill each of the technical and management leadership positions.

From there, you have four sources of candidates to staff them:

1. Team members who are ready to fill the roles now.
2. Team members who can grow into the roles in the time frame.
3. Internal transfers from within your company.
4. External hires who already have the skills.

That is probably an *ordered* list of how you should try to fill the key roles. This is true both because you want people who already know your culture and because reorganizations that depend on yet-to-be-hired individuals are much harder to pull off successfully.

Specifically, I'd recommend having a spreadsheet listing every single person's name, their current team, and their future team. Accidentally missing someone is the cardinal sin of reorganization.

3.7.6 Commit to moving forward

Now it's time to make a go decision. A few questions to ask yourself before you decide to fully commit:

1. Are the changes meaningful net positive?
2. Will the new structure last at least six months?
3. What problems did you discover during design?
4. What will trigger the reorg *after* this one?
5. Who is going to be impacted most?

After you've answered those questions, make sure to get not only your own buy-in but also buy-ins from your peers and leadership. Organizational change is rather resistant to rollback, so you have to be collectively committed to moving forward with it, even if it runs into challenges along the way (which, if history holds, it almost certainly will).

3.7.7 Roll out the change

The final, and oftentimes most awkward, phase of a reorganization is its rollout. There are three key elements to a good rollout:

1. Explanation of reasoning driving the reorganization.
2. Documentation of how each person and team will be impacted.
3. Availability and empathy to help bleed off frustration from impacted individuals.

In general, the actual tactics for doing this are:

1. Discuss with heavily impacted individuals in private first.
2. Ensure that managers and other key individuals are prepared to explain the reasoning behind the changes.
3. Send an email out documenting the changes.
4. Be available for discussion.
5. If necessary, hold an organization all-hands, but probably try not to. People don't process well in large groups, and the best discussions take place in small rooms.
6. Double down on doing skip-level one-on-ones.

And with that, you're done! You've worked through an engineering reorganization. Hopefully, you won't need to do that again for a while.

As a closing thought, an organization is both (1) a collection of people and (2) a manifestation of an idea separate from the individuals comprising it. You can't reason about organizations purely from either direction. There are many, exceedingly valid, different ways to think about any given reorganization, and you should use these ideas as *one* model for thinking through changes, not as a definitive roadmap.

3.8 Identify your controls

When I transitioned from directly supporting teams to instead partnering their managers, I struggled to remain effective without understanding their day-to-day tasks. My first instincts were to retain the same fidelity of context over a much wider area, and for the individuals working with me this was probably indistinguishable from micromanagement. Maybe it even *was* micromanagement.

Thanks to a great deal of feedback and reflection, I've gotten more deliberate at identifying where to engage and where to hang back, a process that I call *identifying your controls*.

Controls are the mechanisms that you use to align with other leaders you work with, and they can range from defining metrics to sprint planning (although I wouldn't recommend the latter). There is no universal set of controls—depending on the size of team and your relationships with its leaders, you'll want to mix and match—but the controls structure itself is universally applicable.

Some of the most common controls that I've seen and used:

Metrics[26] align on outcomes while leaving flexibility around how the outcomes are achieved.

Visions[27] ensure that you agree on long-term direction while preserving short-term flexibility.

Strategies[28] confirm you have a shared understanding of the current constraints and how to address them.

Organization design allows you to coordinate the evolution of a wider organization within the context of sub-organizations.

Head count and transfers are the ultimate form of prioritization, and a good forum for validating how organizational priorities align across individual teams.

Roadmaps align on problem selection and solution validation.

Performance reviews coordinate culture and recognition.

Etc. There are an infinite number of other possibilities, many of which are specific to your company's particular meetings and forums. Start with this list, but don't stick to it!

For whatever controls you pick, the second step is to agree on the *degree of alignment* for each one. Some of the levels that I've found useful are:

I'll do it. Stuff that I will personally be responsible for doing. When you're going to do something, it's better to be explicit and avoid confusion on responsibilities. Best used sparingly.

Preview. I'd like to be involved early and often. This is probably an area where we aren't quite on the same page and this will help us avoid redoing work. and this helps us avoid redoing work.

Review. I'd like to weigh in before it gets published or fully rolled out, but we're pretty aligned on the topic.

Notes. Projects I'd like to follow but don't have much I can add to. Often used for wide-reaching initiatives on which we are well aligned, and I want to be able to represent my colleagues' work correctly.

No surprises. The work that we're currently aligned on but requires updates to keep my mental model in tact. If I'm asked about a related problem, I want to be able to answer it correctly. This is particularly important for me, as my effectiveness is evaluated based on my ability to stay on top of new problems.

Let me know. We're well aligned on this, since my colleagues have done it before and done it well. I want them to let me know if something comes up that I can help with, but otherwise I'm totally confident it'll go well, so we don't need to talk about this much.

Combine your controls and the degree of alignment for each, and you've established the interface between you and the folks you support. This reduces the ambiguity of how you work together and allows everyone to focus. It's useful for agreeing on performance goals, and is also very useful for exposing alignment gaps between you and ıdividuals you work with (For example, it's a worrisome sign if you want to preview every bit of someone's work, unless you just started working together.)

Finally, this is a useful diagnostic for you as a leader to identify if you are micromanaging. If you simply can't imagine a world where you don't preview everyone's work, it's probably time to reflect a bit on what's holding you back from letting the team thrive.

3.9 Career narratives

A peculiar challenge of management is trying to invest in someone's career development when they themselves are uncertain about their goals. As a manager, you may have more experience and more access to opportunities within the company, but that represents a small slice of someone's career possibilities. Our schooling often rewards us for being methodical, linear thinkers, but that approach is less effective outside the intentionally constrained possibility spaces.

The options for approaching a career, particularly for those of us privileged to work in technology, are so extraordinarily vast that exploring them effectively requires a different approach. This vastness also means that you, as a manager, can't simply give folks a career path: you'll inevitably steer them toward the most obvious avenues and through avoidable competition.

Flipping perspectives, it's also quite challenging to plan your own career. I sometimes find myself walking from one meeting where I'm coaching someone on their career goals, straight into a second meeting where I struggle to string together words to articulate my own. The hardest bit is that most folks are always at the furthest point in their career, each change a step into the unknown, with limited visibility into the upcoming opportunities that their company can provide.

The intersection that I've found between the individual's and their manager's perspective is the *career narrative*. I've explained these fabled documents a few times before, in "Roles over Rocket Ships"[29] and "Partnering with Your Manager,"[30] but given how useful they can be, it's useful to expand a bit on the process of maintaining one.

3.9.1 Artificial competition

If you took 10 minutes to ask a dozen people about their immediate career goals, I suspect that for 11 of them it would center on either getting promoted or switching companies to reach the next evolution of their current job. This doesn't mean that climbing the career ladder is bad—that's what it's designed for—but it has the side effect of funneling most folks toward a constrained pool of opportunity.

What I've slowly but increasingly come to believe is that there is much more opportunity on career ladders than off of them, and by including those opportunities you'll make and *feel* more progress. Better yet, you'll find far more opportunities to partner with your peers, no longer competing for limited promotion slots.

For example, if your long-term goal is to be the head of engineering at a mid size company, you could approach that linearly by trying to expand your role bit by bit at your current company on the track to becoming its head of engineering. That'll work for roughly one person at the company, but for everyone else pursuing that same path it will probably be suboptimal.

A different approach would be to instead work on identifying the gaps that would keep you from being a strong head of engineering, and then start using your current role to help fill those gaps. A prototypical head of engineering will be skilled at organizational design, process design, business strategy, recruiting, mentoring, coaching, public speaking, and written communication. They'll also have a broad personal network and a broad foundation from product engineering to infrastructure engineering. That's not even a particularly complete list of relevant skills! There are so many different aspects to build out, and you can find opportunities to practice all of them in your current role. There's no need to convince yourself that your current role is holding you back—everything you need is here.

Importantly, while generalized career paths won't necessarily align cleanly with your goals, they are also unlikely to take full advantage of your strengths. An important part of setting your goals is developing areas you're less experienced in to maximize your global success, but it's equally important to succeed locally within your current environment by prioritizing doing what you do well.

With all of this in mind, take an hour and write up as many goals as you can for what you'd like to accomplish in the next one to five years. Then prioritize the list, pick a few that you'd like to focus on for the next three to six months, and share it with your manager at your next one-on-one.

3.9.2 Translating goals

Once you've identified goals to pursue, the next step is to translate those goals into actions, and this is where your manager can be a leveraged partner in iterating on your career narrative.

Managers tend to have a strong sense of the business's needs, and that gives them the superpower of finding the intersection of your interests and the business's priorities. That translation is a creative pursuit, so don't leave this entirely to your manager: participate as well! Brainstorm projects, research how folks at other companies have pursued similar goals, and educate your manager on aspects of your goals that they don't know much about. (For example, engineers often have more conference speaking experience than engineering managers do.)

Bringing your list of goals to this discussion helps ensure that it's successful. If you don't bring a rough draft, by default you'll get steered toward the contested commons, and your career narrative will be a dull instrument for progress.

This refined list of goals, aligned to your company's priorities, then becomes a central artifact for how you and your manager collaborate on your career evolution. Every quarter or so, take some time to refresh the document and review it together.

If you're unconvinced that it's worth your time to write a career narrative, you certainly don't have to write one. Most folks get away with not writing one for their entire career and have deeply fulfilling careers despite the absence.

That said, if you don't, then there is probably no one guiding your career. Chasing the next promotion is at best a marker on a mass-produced treasure map, with every shovel and metal detector re-covering the same patch. Don't go there. Go somewhere that's disproportionately valuable to you because of who you are and what you want.

3.10 The briefest of media trainings

When I was working at Digg,[31] I was fortunate enough to get five minutes of media training from my colleague Christine. As a testament to her, that brief training lodged deeply in my head, and I've found myself repeating it frequently ever since. Eventually, I realized that I should probably just write it up!

The three rules for speaking with the media:

1. **Answer the question you want to be asked.** If someone asks a very difficult or challenging question, reframe it into one that you're comfortable answering. Don't accept a question's implicit framing, but instead take the opportunity to frame it yourself. *Don't Think of An Elephant* by George Lakoff [32] is a phenomenal, compact guide to framing issues.

2. **Stay positive.** Negative stories can be very compelling. They are quite risky, too! As an interviewee, find a positive framing and stick to it. This is especially true when it comes to competitors and controversy.

3. **Speak in threes.** Narrow your message down to three concise points, make them your refrain, and continue to refer back to your three speaking points.

That was it! Concise, compact, and I'm still using and learning from that advice a decade later.

3.11 Model, document, and share

Early on in my career, I spent a lot of time trying to find *my* leadership style. Recently, I think it's more useful to think about growing yourself as a leader by developing a range of styles and applying them thoughtfully to your circumstances. Confining yourself to one style is just too hard.

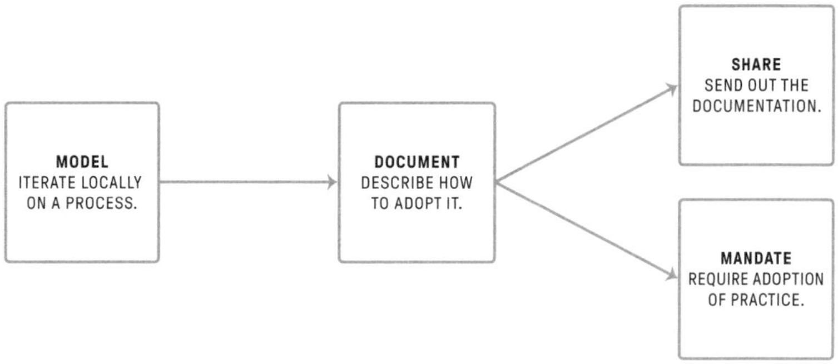

Figure 3.7
The Model, Document, Share-Mandate approach.

One of the trickiest, and most common, leadership scenarios is leading without authority, and I've written about one of the styles that I've found surprisingly effective in those conditions. I call it **Model, Document, Share.**

3.11.1 How it works

Imagine that you've started a new job as an engineering manager, and the teams around you are too busy to use a planning process. You've mentioned to your peers a few times that you've seen kanban[33] work effectively, but folks tried it two years ago and are still upset whenever the word is mentioned: it just doesn't work here.

Your first reaction might be to confront this head-on, but it takes a while to build credibility after starting a new job. Sure, you've been hired for your experience, so they respect your judgment, but it's a hard sell to convince someone that your personal experience should invalidate their personal experience.

I've been trying something different.

Model. Start measuring your team's health (maybe using short, monthly surveys) and your team's throughput (do some lightweight form of task sizing, even if you just do it informally with a senior engineer on the team or with yourself), which will allow you to establish the baseline before your change.

Then just start running kanban. Don't publicize it, don't make a big deal about it, just start doing it with your team. Frame it as a short experiment with the team, and start trying it. Keep iterating on it until you're confident it works. Have the courage to keep at it for a while, and also the courage to stop doing it if it doesn't work after a month or two. Use the team's health and throughput metrics to ground your decision about whether it's working.

Document. After you've discovered an effective approach, document the problem you set out to solve, the learning process you went through, and the details of how another team would adopt the practice for themselves. Be as detailed as possible: make a canonical document, and even get a few folks on other teams to check that it's readable from their perspective.

Share. The final step is to share your documented approach, along with your experience doing the rollout, in a short email. Don't ask everyone to adopt the practice, don't lobby for change, just present the approach and your experience following it.

You'll spend the majority of your time refining approaches that work effectively for your team, a bit of your time documenting how you did it, and almost no time trying to convince folks to change their approach.

Strangely, in my experience, this has often led to more adoption than top-down mandates have.

3.11.2 Where it works

When considering how the **Model, Document, Share** approach works, it's interesting to compare it with the top-down mandate.

Mandates assume:
It's better to adopt a good-enough approach quickly.

- Teams have the bandwidth to adopt a new approach.
- The organization has available resources to coordinate a rollout.
- You want to centralize decision-making on this topic.

- Consistency is important; all teams need to approach this problem in the same way.
- It's important to make this change quickly.

Model, Document, Share assumes:
It's better to adopt a great approach slowly.

- Some teams don't have the bandwidth to adopt a new approach.
- The organization may not have resources to coordinate a rollout.
- You want to decentralize decision-making on this topic.
- Teams have agency to adopt the appropriate practices for themselves.
- It's okay to approach change gradually.

If your circumstances and your organization's values align with the second list, then this approach may be *more* effective for you than making mandates. If you have the time, you can slowly flock toward great practice, without needing organizational authority. (You'll still need some natural authority, the respect of your peers.)

Although I've seen this approach work remarkably well, I've also seen it go nowhere. It's a particular tool for certain circumstances, and it does fail. It may be an inexpensive failure—folks simply don't adopt—as you haven't spent much authority on it, but nonetheless you still haven't accomplished your goal. It's particularly important not to try to use this as a strategy to circumvent organizational authority. Operating in direct conflict with authority usually doesn't end very well.

3.12 Scaling consistency: designing centralized decision-making groups

In small organizations, it's easy for individuals to be aware of what others are doing and to remember how they've previously approached similar problems. This hive mind and memory create decision-making whose consistency correlates strongly with quality. As organizations grow, there is a subtle slide into inconsistency, which is often one of the most challenging aspects of evolving from a small team into a much larger one.

There are many different approaches to try to manage inconsistency creep. Some of the solutions I've seen are formalized sprints, training, shadowing, documentation, code linters,[34] process automation (particularly deploys), and incident reviews. However, when the problem becomes truly acute, folks eventually reach for the same tool: adding a centralized, accountable group.

The two most common flavors of this I've seen are "product reviews" to standardize product decisions and the "architecture group" to encourage consistent technical design. There are hundreds of varieties, and they crop up wherever decisions are made.

Some of these groups take on an authoritative bent, becoming rigid gatekeepers, and others become more advisory, with a focus on educating folks toward consistency. Depending on your culture and how you value consistency, there are an infinite number of approaches, and designing an effective decision-making group depends on a handful of core decisions.

3.12.1 Positive and negative freedoms

Before jumping into design, a few words on the framing that I use for reasoning about when to create a new centralized authority.

These groups typically consolidate significant authority from the broader community into the hands of a few. Many folks will feel a significant loss of freedom when you create these groups, as their zone of decision-making will be newly limited. Less obviously, many folks find the creation of centralized groups to be extremely empowering. The difference? One group is largely populated by individuals comfortable with self-authorization, and the latter typically have a higher threshold for self-authorization.

This is just one example of the dynamics that play out across many dimensions when you're considering introducing a new authority, and the most useful framework I've found for thinking through this involves *positive* and *negative freedoms*. A *positive freedom* is the freedom to do something, for example the freedom to pick a programming language you prefer. A *negative freedom* is the freedom from things happening to you, for example the freedom not to be obligated to support additional programming languages, even if others would greatly prefer them.

How are you shifting freedoms, and who are you shifting them from? Particularly in cases where ownership is extremely diffuse, I believe that cautiously authoritative groups do increase net positive freedom for individuals without greatly reducing negative freedom. That also happens to be my goal when designing a new group!

3.12.2 Group design

Now that you've decided to create a decision-making group, it's time to get into the choices!

Influence. How do you expect this group to influence results? Will they be an authoritative group that makes binding decisions? Will you rely on the natural authority of the members you select? Will they primarily work through advocacy? The answers to these questions along with the particular folks in the group, will be the primary factor in how your group impacts the positive and negative freedoms of those they work with.

Interface. How will other teams interact with this team? Will they submit tickets, send emails, attend a weekly review session? Will you be reviewing work before it launches, or previewing designs before they're staffed? Depending on the kind of influence they're exerting and how your company works, you'll want to play around with different approaches.

Size. How large should the group be? If it's six or fewer individuals, it's possible for them to gel into a true team, one whose members know each other well, work together closely, and shift a significant portion of their individual identities into the team. If the group has more than ten members, you'll find it hard to even have a good discussion, and it'll need to be structured into sub-groups to function well (rotation that spreads members over time, working groups of some members to focus on particular topics, and so on). The larger the group, the more responsibility becomes diffuse, and the more you'll need to have specified roles within the group, for example a member responsible for coordinating members' focus.

Time commitment. How much time will members spend working in this group? Will this be their top priority, or will they still primarily be working on other projects? Higher time commitment correlates with more actions and decisions. Consequently, my sense is that you want time commitment to be higher for areas where folks are directly impacted by the consequences of their decisions, and to be lower for scenarios with weaker feedback loops.

Identity. Should members view their role in the group as their primary identity? Should they continue to identify primarily as members of their existing team? You'll need a small team and high time commitment to support individuals shifting their identity.

Selection process. How will you select members? I've found the best method to be a structured selection process,[35] in which you identify the requirements to be a member and the skills that you believe will be valuable, and then allow folks to apply. Membership in these groups often becomes an important signal of organizational status, which makes having a consistent process for selecting membership especially important.

Length of term. How long will members serve? Are these permanent assignments, or are they fixed terms for, say, six months? If they are fixed terms, are folks eligible for subsequent elections? Is there a term limit? I've tried most combinations, and my sense is that the best default is fixed terms, while allowing current individuals to remain eligible, and without enacting term limits.

Representation. How representative will this group be? Will you explicitly select folks based on their teams, tenure, or seniority, or will you allow clusters? Attention to this can help you avoid architecture reviews that are missing front-end and product engineers, as well as product reviews without infrastructure perspective.

Predictably, each of these decisions will impact the effectiveness of the others, which can make designing the group you want quite tricky. Some formats will require particular kinds of individuals to staff them, and you have to design groups that will work with the people available to participate and the culture they are participating within.

3.12.3 Failure modes

Before you release that email you're writing to spin up a new centralized decision-making group, it's worth talking about the four ways these groups consistently fail. They tend to be domineering, bottlenecked, status-oriented, or inert.

Domineering groups significantly reduce individuals' negative and positive freedoms, and become churn factories for members. This is most common when those making decisions are abstracted from the consequences of the decisions, e.g., architecture groups in which the members write little code.

Bottlenecked groups tend to be very helpful, but are trying to do more then they're actually able to do. These groups get worn down, and either burn out their members or create a structured backlog to avoid burning themselves out, which ends up seriously slowing down folks who need their authority.

Status-oriented groups place more emphasis on being a member of the group than on the group's nominal purpose. The value of the group revolves around recognition rather than contribution, leading to folks who try to join the group for status, and the diffusion of whatever original mission the group set out to resolve.

Inert groups just don't do much of anything. Typically, these are groups whose members have not gelled or are too busy. On the plus side, this is by far the most benign of the four failure modes—but you're also missing out on a great deal of opportunity!

Having experienced each of these a few times, I try to ensure that there is a manager embedded into every centralized group, and that the manager is responsible for iterating on the format to dodge these pitfalls.

3.13 Presenting to senior leadership

Yahoo! BOSS[36] was partially powered by an internal Yahoo! search technology named Vespa. We'd run into a bunch of challenges, and I'd decided to convince my team we should migrate to SOLR.[37] My manager asked me to put together a presentation for our next team meeting. The meeting came, I started to present, and within two slides things fell apart.

"No, no, this isn't the way to put this together. This is like an academic presentation. You have to start from the conclusion first," my manager lamented as he abruptly terminated my presentation. Pausing to brush my deck's fragments from his boots, he offered some final wisdom: "And don't use curved lines in your diagrams. Those *never* make sense."

It took me a few years to glean a lesson from that experience, but it comes to mind frequently now when I work with folks who are just starting to present to executives. Giving a presentation to senior leadership is a bit of a dark art: it takes a while to master, and most people who do it well can't quite articulate how they do it. Worse yet, many people who are excellent rely on advantages that resist replication: charisma, quick wit, deep subject matter expertise, or years of experience.

That said, few people watching me bomb my Yahoo! presentation would have bet I'd ever figure this one out, so you should know that it is a learnable skill. Along the way, I've picked up some tips that I hope will help you prepare for your next presentation:

Communication is company-specific. Every company has different communication styles and patterns. Successful presenters probably can't tell you what they do to succeed, but if you watch them and take notes, you'll identify some consistent patterns. Each time you watch individuals present to leadership, study their approach.

Start with the conclusion. Particularly in written communication, folks skim until they get bored and then stop reading. Accommodate this behavior by starting with what's important, instead of building toward it gradually.

Frame why the topic matters. Typically, you'll be presenting on an area that you're intimately familiar with, and it's probably very obvious to you why the work matters. This will be much less obvious to folks who don't think about the area as often. Start by explaining why your work matters to the company.

Everyone loves a narrative. Another aspect of framing the topic is providing a narrative of where things are, how you got here, and where you're going now. This should be a sentence or two along the lines of, "Last year, we had trouble closing several important customers due to concerns about our scalability. We identified our databases as our constraints to scaling, and since then our focus has been moving to a new sharding model that enables horizontal scaling. That's going well, and we expect to finish in Q3."

Prepare for detours. Many forums will allow you to lead your presentation according to plan, but that is an unreliable prediction when presenting to senior leadership. Instead, you need to be prepared to

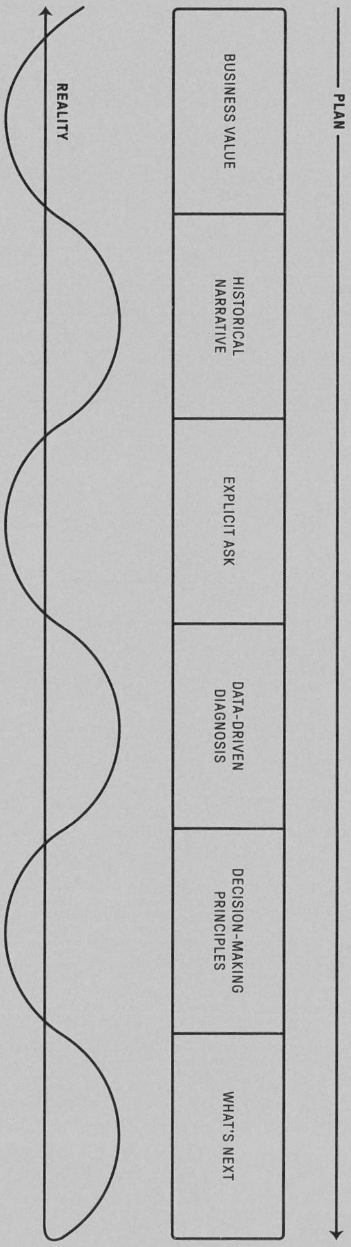

Figure 3.8
The expected and actual experience of presenting to executives.

lead the entire presentation yourself, while being equally ready for the discussion to derail toward a thread of unexpected questions.

Answer directly. Senior leaders tend to be indirectly responsible for wide areas, and frequently pierce into areas to debug problems. Their experience "debug piercing" tunes their radar for evasive answers, and you don't want to be a blip on that screen. Instead of hiding problems, use them as an opportunity to explain your plans to address them.

Deep in the data. You should be deep enough in your data that you can use it to answer unexpected questions. This means spending time exploring the data, and the most common way to do that is to run a thorough goal-setting exercise.[38]

Derive actions from principles. One of your aims is to provide a mental model of how you view the topic, allowing folks to get familiar with how you make decisions. Showing you are "in the data" is part of this. The other aspect is defining the guiding principles you're using to approach decisions.

Discuss the details. Some executives test presenters by diving into the details, trying to uncover an area the presenter is uncomfortable speaking on. You should be familiar with the details, e.g., project statuses, but I think that it's usually best to reframe the discussion when you get too far into the details. Try to pop up to either the data or the principles, which tend to be more useful conversations.

Prepare a lot; practice a little. If you're presenting to your entire company, practicing your presentation is time well spent. Leadership presentations tend to quickly detour, so practice isn't quite as useful. Practice until you're comfortable, but prefer to prepare instead getting deeper into the data, details, and principles. As a corollary, if you're knowledgeable in the area you're responsible for, and have spent time getting comfortable with the format, over time you'll find that you won't need to prepare much for these specifically. Rather, whether you're able to present effectively without much preparation will become a signal for whether you're keeping up with your span of responsibility.

Make a clear ask. If you don't go into a meeting with leadership with a clear goal, your meeting will either go nowhere or go poorly. Start the meeting by explicitly framing your goal!

That's a lot to remember, so I've synthesized these ideas into a loose template. There absolutely is not a single right way to present to senior leaders, but hopefully the template is a useful starting point.

My general approach to presenting to senior leaders is:

1. **Tie topic to business value.** One or two sentences to answer the question "Why should anyone care?"

2. **Establish historical narrative.** Two to four sentences to help folks understand how things are going, how we got here, and what the next planned step is.

3. **Explicit ask.** What are you looking for from the audience? One or two sentences.

4. **Data-driven diagnosis.** Along the lines of a strategy's diagnosis phase,[39] explain the current constraints and context, primarily through data. Try to provide enough raw data to allow people to follow your analysis. If you only provide analysis, then you're asking folks to take you on trust, which can come across as evasive. This should be a few paragraphs, up to a page.

5. **Decision-making principles.** Explain the principles that you're applying against the diagnosis, articulating the mental model you are using to make decisions.

6. **What's next and when it'll be done.** Apply your principles to the diagnosis to generate the next steps. It should be clear to folks reading along how your actions derive from your principles and the data. If it's not, then either rework your principles or your actions!

7. **Return to explicit ask.** The final step is to return to your explicit ask and ensure that you get the information or guidance you need.

I've had a lot of luck with this format in general, and I think you'll find it pretty useful as a starting point. That said, the first rule remains true: communication is company-specific. If things don't quite work for this format in your company, then watch how other folks present. Given a few examples, you'll be able to reverse engineer the discussions that go well into a workable template.

3.14 Time management

When you sit down for coffee with a manager, you can probably guess the biggest challenge on their mind: time management. Sure, time management isn't *always* everyone's biggest challenge, but once the crises of the day recede, it comes to the fore.

Time management is the enduring meta-problem of leadership. For most other aspects of leadership, you can look to more experienced managers and be reassured that things will get better, but in this dimension it appears that the most tenured folks are the ones most underwater. Yes, their degree of difficulty is certainly higher, but it's intimidating to consider that there's little evidence most folks ever get a solid grasp of their time.

Does that make this a lost cause? Nah.

I'm still pretty busy on a day-to-day basis, but I've gotten much, much better at getting things done, not by getting faster but by getting more logical about solving problems. The most impactful changes I've made to how I manage time are:

Quarterly time retrospective. Every quarter, I spend a few hours categorizing my calendar from the past three months to figure out how I've invested my time. This is useful for me to reflect on the major projects I've done, and also to get a sense of my general allocation of time. I then use this analysis to shuffle my goal time allocation for the next quarter.

Most folks are skeptical of whether this is time well spent, but I've found it particularly helpful, and it's the cornerstone of my efforts to be mindful of my time.

Prioritize long-term success over short-term quality. As your scope increases, the important work that you're responsible for may simply not be possible to finish. Worse, the work that you believe is most important, perhaps high-quality one-on-ones, is often competing with work that's essential to long-term success, like hiring for a critical role. Ultimately, you have to prioritize long-term success, even if it's personally unrewarding to do so in the short term. It's not that I like this approach, it's that nothing else works.

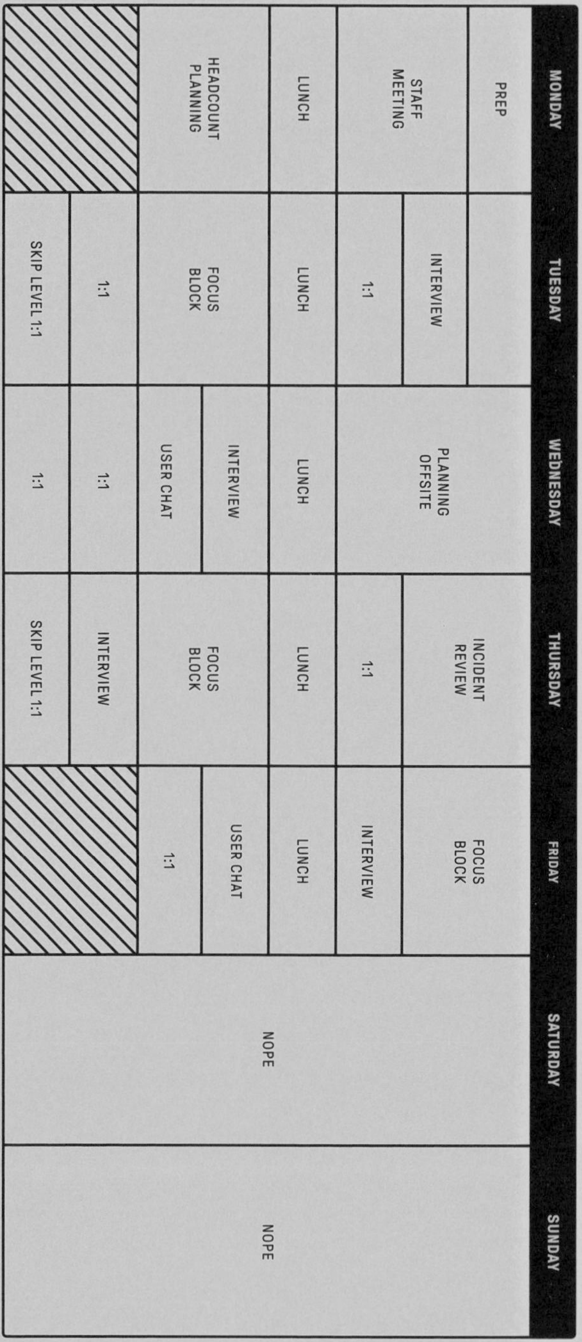
MONDAY
PREP
STAFF MEETING
LUNCH
HEADCOUNT PLANNING
TUESDAY
INTERVIEW
1:1
LUNCH
FOCUS BLOCK
1:1
SKIP LEVEL 1:1
WEDNESDAY
PLANNING OFFSITE
LUNCH
INTERVIEW
USER CHAT
1:1
1:1
THURSDAY
INCIDENT REVIEW
1:1
LUNCH
FOCUS BLOCK
INTERVIEW
SKIP LEVEL 1:1
FRIDAY
FOCUS BLOCK
INTERVIEW
LUNCH
USER CHAT
1:1
SATURDAY
NOPE
SUNDAY
NOPE

Figure 3.9
Calendar time-blocking for an engineering manager.

Finish small, leveraged things. If you're doing leveraged work,[40] then each thing that you finish[41] should create more bandwidth to do more future work. It's also very rewarding to finish things. Together, these factors allow large volumes of quick things to build into crescendoing momentum.

Stop doing things. When you're quite underwater, a surprisingly underutilized technique is to stop doing things. If you drop things in an unstructured way, this goes very poorly, but done with structure this works every time. Identify some critical work that you won't do, recategorize that newly unstaffed work as organizational risk,[42] and then alert your team and management chain that you won't be doing it. This last bit is essential: it's fine to drop things, but it's quite bad to silently drop them.

Size backward, not forward. A good example of this is scheduling skip-levels.[43] When you start managing a multi-tier team, say 20 individuals, you can specify a frequency for skip-levels and reason forward to figure out how many hours of skip-levels you'll do in a given week. Say you have 16 indirect reports, and you want to see them once a month for 30 minutes, so you end up doing two hours per week.

This stops working as your team grows, because there is simply no reasonable frequency that won't end up consuming an unsustainable number of hours. Instead, specify the number of hours you're able to dedicate to the activity, perhaps two per week, and perform as many skip-levels as possible within that amount of time. This method keeps you in control of your time allocation, and it scales as your team grows.

Delegate working "in the system." Wherever you're working "in the system,"[44] design a path that will enable someone else to take on that work. It might be that this plan will take a year to come together, and that's fine, but what's not all right is if it's going to take a year and you haven't even started.

Trust the system you build. Once you've built the system, at some point you have to learn to trust it. The most important case of this is handing off the responsibility to handle exceptions. Many managers hold onto the authority to handle exceptions for too long, and at that point you lose much of the system's leverage. Handling exceptions can easily consume all of your energy, and either delegating them or designing them out of the system is essential to scaling your time.

Decouple participation from productivity. As you grow more senior, you'll be invited to more meetings, and many of those meetings will come with significant status. Attending those meetings can make you feel powerful, but you have to keep perspective about whether you're accomplishing much by attending. Sometimes, being able to convey important context to your team is super valuable, and in those cases you should keep attending, but don't fall into the trap of assuming that attendance is valuable.

Hire until you are slightly ahead of growth. The best gift of time management that you can give yourself is hiring capable folks, and hiring them before you get overwhelmed. By having a clear organizational design, you can hire folks into roles before their absence becomes crippling.

Calendar blocking. Creating blocks of time on your calendar is the perennial trick of time management: add three or four two-hour blocks scattered across your week to support more focused work. It's not especially effective, but it does work to some extent and is quick to set up, which has made me a devoted user.

Getting administrative support. Once you've exhausted all the above tools and approaches, the final thing to consider is getting administrative support. I was once quite skeptical of whether admin support is necessary—and, until your organization and commitments reach a certain level of complexity, it isn't—but at some point having someone else handling the dozens of little interruptions is a remarkable improvement.

As you start using more of these approaches, you won't immediately find yourself less busy, but you will gradually start to finish more work. Over a longer period of time, though, you can get less busy by prioritizing finishing things with the goal of reducing load. If you're creative and consequent, and if you don't fall into the trap of believing that being busy is being productive, you'll find a way to get the workload under control.

3.15 Communities of learning

I've always preferred learning in private. Got something difficult? Sure, leave me alone for a few hours and I can probably figure it out. If you want me to figure it out with you watching, I'm not even sure how to start. This is partly introversion, but altogether I'm pretty uncomfortable making mistakes in public. Like a lot of folks, I have a brain that still helpfully reminds me of public errors I made decades ago, and they still bother me.

For a very long time, this discomfort prevented me from discovering one of the most rewarding elements of being in a supportive work environment: building a community of learning with your peers. This works especially well in a gelled "first team,"[45] and, recently I've been spending more time facilitating a broader learning community of engineering managers.

When I first started facilitating the group, we focused on content-rich presentations. Each slide was dense with important lessons and essential details. It didn't work well. Folks weren't engaged. Attendance dropped. Learning was not the order of the day.

Since then, we've iterated on format and eventually stumbled on an approach that has worked consistently:

1. **Be a facilitator, not a lecturer.** Folks want to learn from each other more than they want to learn from a single presenter. Step back and facilitate.

2. **Brief presentations, long discussions.** Present a few minutes of content, maybe five, and then move into discussion. Keep the discussions short enough that even if a group doesn't get traction on a given topic, it doesn't become awkward. A good time limit would be 10 minutes.

3. **Small breakout groups.** Giving folks time to discuss in small groups allows them to learn a bit about the topic in a small, safe place. It also gives everyone an opportunity to be part of the discussion, which is a lot more engaging than listening to others for an hour.

4. **Bring learnings to the full group.** After discussions, give each group an opportunity to bring their discussion back to the larger group, to allow the groups to cross-pollinate their learnings.

5. **Choose topics that people already know about.** Successful topics are ones that people have already thought about, typically because these concepts are core to their daily work. Ideally the topic is something that they do but would like to get better at, such as one-on-ones, mentorship, coaching, or career development.

COMMUNITIES OF LEARNING		
	CHECK IN	SHARE: YOUR NAME, YOUR TEAM, 1 SENTENCE ON TOP PRIORITY.
	OPPORTUNITIES	A LOT OF LEARNING IS TOP-DOWN. WE CAN LEARN A LOT FROM PEERS. LEARNING AS A HABIT, LEARNING IN A COMMUNITY.
	DISCUSSION	BREAK INTO GROUPS OF 4–5. WHERE DO YOU LEARN TODAY? WHO DO YOU LEARN FROM? REGROUP IN 10 MINUTES.
	REGROUP	SUMMARY FROM EACH GROUP.

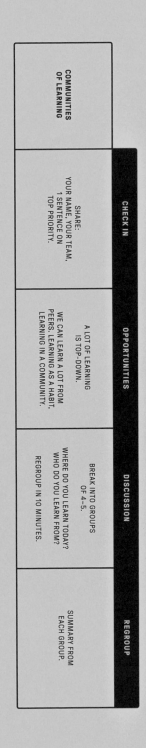

Figure 3.10
Structuring a meeting to facilitate a community of learning.

People find it very hard to discuss content that they've just learned if it's too far away from their previous experience. That also creates an environment where learning has to come from the facilitator instead of from the group at large.

6. **Encourage tenured folks to attend.** For many learning communities, you'll find that the most senior or most tenured folks opt out to focus on other work. This is a shame, because there is so much for them to teach newer folks, and also because it creates an opportunity for them to learn from and get to know new members.

7. **Optional pre-reads.** Some folks aren't comfortable being introduced to a new topic in public, and for those individuals, providing a list of optional pre-reads can help them prepare for the discussion. I find that most people don't read them (which, surprisingly, I also found true when hosting paper-reading groups[46]), but for some folks they're very helpful.

8. **Checking in.** Depending on the size of the group, it can be powerful to start by checking in with each other, having each person give a 20- or 30-second self-introduction. The format we've been using lately is your name, your team, and one sentence about what's on your mind. This is especially useful in quickly growing communities, as it makes it easier for individuals to meet each other.

   The thing I enjoy most about this format is that it gives folks what they really want, spending time learning from each other, and is remarkably quick for the facilitator to prepare. I'm far from a seasoned facilitator, and I've also found this format to be a rewarding and safe opportunity for me to grow as a facilitator.

   If your company doesn't have any learning communities, give it a try. I've found this one of the easiest, most rewarding things I've had the opportunity to work on.

4

CHAPTER 4

Approaches

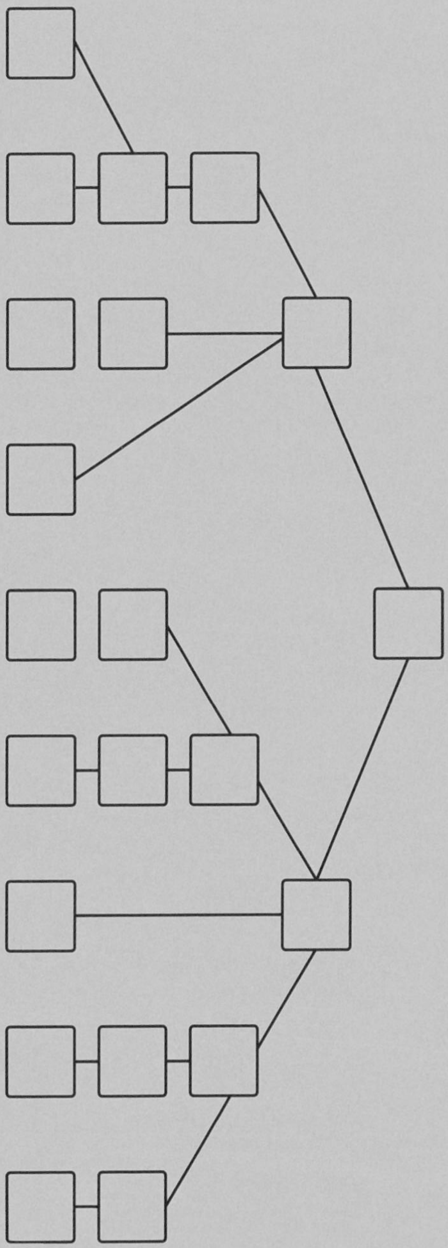

Figure 4.1
Visual representation of a well and poorly balanced organization.

Approaches

You unravel most puzzles knowing they're solvable. You play most games guided by a rule book. For engineering managers, challenges emerge unexpected from a hundred small decisions, with few rules and no promises. Many of these challenges are difficult in the worst sense. There are few options for next steps and pursuing any one of them feels problematic.

This chapter covers several such problems, including handling policy exceptions, pushing back on requests from your management chain, and scaling yourself in times of transition. Management is an ethical profession, and our decisions matter, especially the hard ones.

4.1 Work the policy, not the exceptions

At an early job, I worked with a coworker whose philosophy was "If you don't ask for it, you'll never get it." Which culminated in continuous escalations to management for exceptional treatment. This approach was pretty far from my intuition of how a well-run organization should work, and it grated on my belief that consistency is a precondition of fairness.

Since then, I've come to believe that environments that tolerate frequent exceptions are not only susceptible to bias but are also inefficient. Keeping an organization aligned is challenging in the best of times, and exceptions undermine one of the most powerful mechanisms for alignment: consistency.

Conversely, organizations survive by adapting to the dynamic circumstances they live in. An organization stubbornly insisting on its established routines is a company pacing a path whose grooves lead to failure.

How do you reconcile consistency and change?

As with most seemingly opposing goals, the more time I spent considering them, the less they were mutually exclusive. Eventually, a unified approach emerged, which I call "*Work the policy, not the exceptions.*"

4.1.1 Good policy is opinionated

Every policy you write is a small strategy,[1] built by identifying your goals and the constraints that bring actions into alignment with those goals. It's probably easiest to dig into an example.

One of the most interesting policies that I recently worked on was designing who is able to join which teams in a company with both many remote employees and many geographically distributed offices. Our most important goals were:

1. Make every office a first-tier office; there are no second-tier offices.

   A first-tier office must own multiple critical projects, and its work shouldn't be constrained by support from other offices. Furthermore, it must have many individuals physically present who work together closely.

2. Ensure that remote engineers remain an essential, well-supported cohort of the company.

Once we had agreed upon our goals, the next step was to codify some constraints in order to narrow the scope of allowed actions to those that supported our goals. In this case, a (slightly simplified) set of constraints might be:

1. Teams are staffed in, at most, one office. (I say "at most" in order to support the premise of teams composed entirely of remote engineers.)

2. Employees in an office must be members of a team in that office.

3. Remote employees may work on any team.

4. Employees within a 60 minute commute of an office must work from that office.

These are good examples of constraints because they clearly constrain allowed behaviors. You could imagine less opinionated constraints, such as "folks in an office should prefer working on teams within that office," but that would do less to constrain behavior.

If you find yourself writing constraints that don't actually constrain choice, it's useful to check if you're dancing around an unstated goal that's limiting your options. For example, in the above, you might have an unstated goal that an employee pursuing their preferred work is more important than offices being first-tier, which would lead you toward weak constraint.

The fixed cost of creating and maintaining a policy is high enough that I generally don't recommend writing policies that do little to constrain behavior. In fact, that's a useful definition of bad policy. In such cases, I instead recommend writing norms, which provide nonbinding recommendations. Because they're nonbinding, they don't require escalations to address ambiguities or edge cases.

4.1.2 Exception debt

Once you've supported your goals through constraints, all you have to do is consistently uphold your constraints. This is easy to say, but consistency requires no little bravery. Even with the best intentions, I've often gone astray when it was time for me to support my own policies.

The two reasons that applying policy consistently is particularly challenging are:

1. **Accepting reduced opportunity space.** Good constraints make trade-offs that deliberately narrow your opportunity space. Some of the opportunities that you'll encounter within that space will be exceptionally good, and it's hard to stay true when faced with concrete consequences.

2. **Locally suboptimal.** Satisfying global constraints inevitably leads to local inefficiency, sometimes forcing some teams to deal with deeply challenging circumstances in order to support a broader goal that they may experience little benefit from. It's hard to ask folks to accept such circumstances, harder to be someone in one of those local inefficiencies, and hardest yet to stick to the decisions at real personal cost to the folks you're impacting.

When we've picked thoughtful constraints to allow us to accomplish important goals, we need the courage to bypass these opportunities and accept these local inefficiencies. If we don't summon and maintain that courage, we incur most of the costs and receive few of the benefits.

Policy success is directly dependent on how we handle requests for exception. Granting exceptions undermines people's sense of fairness, and sets a precedent that undermines future policy. In environments where exceptions become normalized, leaders often find that issuing writs of exception—for policies they themselves have designed—starts to swallow up much of their time. Organizations spending significant time on exceptions are experiencing *exception debt*. The escape is to stop working the exceptions, and instead *work the policy*.

4.1.3 Work the policy

Once you've invested so much time into drafting policy, you have to avoid undermining your work, and yourself, with exceptions. That said, you can't simply ignore escalations and exceptions requests, which often represent incompatibilities between the reality you designed your policy for and the reality you're operating in. Instead, collect every escalation as a test case for reconsidering your constraints.

Once you've collected enough escalations, revisit the constraints that you developed in the original policy, merge in the challenges discovered in applying the policy, and either reaffirm the existing constraints or generate a new series of constraints that handle the escalations more effectively.

This approach is powerful because it creates a release valve for folks who are frustrated with rough edges in your current policies—they're still welcome to escalate—while also ensuring that everyone is operating in a consistent, fair environment; escalations will only be used as inputs for updated policy, not handled in a one-off fashion. The approach also maintains working on policy as a leveraged operation for leadership, avoiding the onerous robes of an exceptions judge.

When you roll out a policy, it's quite helpful to declare a future time when you'll refresh it, which ensures that you'll have the time to fully evaluate your new policy before attempting revision. It's fairly common for folks to modify good, effective policy due to concerns that arise before the policy has had time to show its effect. At a sufficiently high rate of change, policy is indistinguishable from exception.

The next time you're about to dive into fixing a complicated one-off situation, consider taking a step back and documenting the problem but not trying to solve it. Commit to refreshing the policy in a month, and batch all exceptions requests until then. Merge the escalations and your current policy into a new revision. This will save your time, build teams' trust in the system, and move you from working the exceptions to working the policy.

4.2 Saying no

Some years back, I was sitting in a room with my manager, our CTO, and a crisis. An engineer on my team had mishandled two alerts, which had cascaded into plausibly the worst production incident that the company had experienced to date. There were three root causes: alert fatigue, a lack of velocity context for out-of-disk-space alerts, and our reliance on a centralized database with little support for vertical scaling. At that moment, though, we were no longer talking about root causes. We were discussing whether to fire the on-call engineer, and I was saying no.

It was in that era of my career that I came to view management as, at its core, a moral profession. We have the opportunity to create an environment for those around us to be their best, in fair surroundings. For me, that's both an opportunity and an obligation for managers, and saying no in that room with my manager and CTO was, in part, my decision to hold the line on what's right. However, there was a second no in that room, and it's one you'll use routinely even under the best of circumstances. That no is an expression of what is possible for the team you lead to do. I felt that the decision would be wrong, but also that the precedent of firing people for on-call mistakes would irreparably damage the morale of a team who already saw their phone batteries drained before the end of a 12 hour on-call shift.

This no is explaining your team's constraints to folks outside the team, and it's one of the most important activities you undertake as an engineering leader.

4.2.1 Constraints

Folks who communicate a no effectively are not the firmest speakers, nor do they make frequent use of the word itself. They are able to convincingly explain their team's constraints and articulate why the proposed path is either unattainable or undesirable.

Articulating your constraints depends on the particulars of the issue at hand, but I find that two topics are frequent venues of disagreement. The first is **velocity**: Why is this taking so long when it should take a couple of hours? The other is **prioritization**: Why can't you work on this other, more important, project?

Let's dig into how to have those conversations constructively.

4.2.2 Velocity

When folks want you to commit to more work than you believe you can deliver, your goal is to provide a compelling explanation of how your team finishes work. *Finishes* is particularly important, as opposed to *does*, because partial work has no value, and your team's defining constraints are often in the finishing stages. The most effective approach that I've found for explaining your team's delivery process is to build a kanban board[2] describing the steps that work goes through, and documenting who performs which steps. You don't have to switch to using a kanban system, although I've found it very effective for debugging team performance, you just have to populate the board once to expose your current constraints.

Using this board, you'll be able to explain what the current constraints are for execution, and help your team narrow suggestions for improvement down to areas that will actually help. If you don't provide this framework, people tend to start making suggestions everywhere across your process, which at best means many ideas won't reduce load where it's most helpful, and at worst may inadvertently increase load.

You want to focus your team on your core constraint, the single inefficient component that's slowing down your throughput of finished work. Once you've focused the conversation on your core constraint, the next step is explaining what's preventing you from solving for it. At many technology companies this comes down to technical debt or toil. However, the specters of technical debt and toil have been used to shirk so much responsibility that simply naming them tends to be unconvincing.

Instead, you have to translate the problem into something resembling data. If you're following a consistent project management methodology, this can be as easy as explaining how you decide the number of story points for each sprint, along with how that number has trended over time. If not, I've found it useful to borrow the approach of a sampling profiler: for a week, check what your team is working on at a few random moments across the day, and use that as an approximation of how time is being spent.

Once you're able to explain your constraints and how time is being spent, then you're having a useful conversation about whether you can shift time from other behaviors toward your constraints. The final stage comes next, which is the discussion around adding capacity.

There are two ways to add capacity: move existing resources to the team (away from what they're currently doing) or create new resources (typically through hiring). Neither is a panacea, and they are explored in "A Case against Top-Down Global Optimization"[3] and "Productivity in the Age of Hypergrowth,"[4] respectively.

Putting it all together, the best outcome of a discussion on velocity is to identify a reality-based approach that will support your core constraint. The second-best outcome is for folks to agree that you're properly allocated against your constraints and to shift the conversation to prioritization. (Those are the only good outcomes.)

4.2.3 Priorities

Although shifting from a discussion about velocity to one about prioritization is a good outcome, expressing your priorities convincingly can be a difficult, daunting task. I recommend breaking it down into three discrete steps: document all your incoming asks, develop guiding principles for how work is selected, and then share subsets of tasks you've selected based on those guiding principles. Hopefully, documenting your incoming asks is as straightforward as auditing your team's tickets, but it's pretty common for some of the most important asks to be undocumented. What I've found effective is to blend existing planning artifacts (typically quarterly/annual plans) and your tickets into a list, and then test it against your most important stakeholders. Keep asking those who routinely have dependencies on your team, "Does this seem like the right list of asks?" The result will be a fairly accurate artifact.

From there, you have to pick the guiding principles that you'll use for selecting tasks. How you'll do that will depend on your team and your company—infrastructure teams will pick different guides[5] than product teams will, but they'll likely be grounded in your company's top-level plans and will intersect with your team's mission. (The most controversial guides tend to be statements about the value of current work versus future work, for example doing investment work today that will pay off in two years but limit value in the short term. One technique that I've found useful for this particular scenario is specifying quotas for both immediate and long-term work.)

The last step is sitting down with your team and applying your guiding principles to the universe of incoming asks, then finding the subset to prioritize. You'll continuously get more requests to do work, so it's important that this process is lightweight enough that you can rerun it periodically.

Which, it so happens, is exactly what you should do when a stakeholder disagrees with your priorities. Understand their requests, and sit down with them to test their ask against your guiding principles and your currently prioritized work. If their request is more important than your current work, shift priorities at your next planning session. (To limit churn created by shifting priorities, it's useful to wait for the next planning session instead of making these changes immediately. This does mean that you'll need to be refreshing your plan at least monthly.)

4.2.4 Relationships

If you've poured time into explaining both your velocity and priorities but your perspective still isn't resonating, then it's fairly likely that you have a relationship problem to address. In that case, the next step isn't investing more energy in explaining your constraints, but instead working on how you partner with your individual stakeholders.

4.3 Your philosophy of management

For me, the first few years of management were a wild frenzy. Every situation was brand-new, and I puzzled through each decision from the ground up. Over time, I developed some rules of thumb and guidelines, but only the experience of managing managers has truly refined my thoughts on management.

When I started managing, my leadership philosophy was simple:

1. The Golden Rule[6] makes a lot of sense.

2. Give everyone an explicit area of *ownership that* they are responsible for.

3. Reward and status should derive from *finishing* high-quality work.

4. *Lead from the front*, and never ask anyone to do something you wouldn't.

These have served well as a foundation, but by applying them repeatedly over time and circumstance, I've seen them fray on the edge cases. Learning forward, I've started to weave in a number of additional ideas in the vain quest for a unified theory of management.

4.3.1 An ethical profession

I believe that management, at its core, is an ethical profession. To see ourselves, we don't look at the mirror, but rather at how we treat a member of the team who is not succeeding. Not at the mirror, but at our compensation philosophy. Not at the mirror, but at how we pitch the roles to candidates. Whom we promote. How we assign raises. Provide growth opportunities. PTO requests. Working hours.

We have such a huge impact on the people we work with—and especially on the people who work "for" us—and taking responsibility for that impact is fundamental to good management.

This doesn't always mean being your team's best friend. Sometimes it means asking them to make personal sacrifices, letting go of a popular member of the team, or canceling a project the team is excited about. It's remembering that you leave a broad wake, and that your actions have a profound impact on those around you.

4.3.2 Strong relationships > any problem

I believe that almost every internal problem can be traced back to a missing or poor relationship, and that with great relationships it is possible to come together and solve almost anything.

Technical disagreements become learning opportunities for everyone. Setbacks are now a shared experience that offers the opportunity to gel together as a team.

Even with great relationships, there are still real challenges! You have a limited budget for giving raises and can't satisfy everyone. If your customers don't love your product, camaraderie can't pay salaries. Some technical problems are genuinely novel, without an obvious solution, and sometimes the obvious solution is cost-prohibitive.

That said, I try to start debugging problems from the relationship angle, and I find this technique pretty effective.

4.3.3 People over process

A few years back, one of the leaders I worked with told me, "With the right people, any process works, and with the wrong people, no process works."

I've found this to be pretty accurate.

Process is a tool to make it easy to collaborate, and the process that the team enjoys is usually the right process. If your process is failing somehow, it's worth really digging into how it's failing before you start looking for another process to replace it.

As you start homing in on the problem (maybe it's you!), honestly ask yourself if a different process would address it, or if you're moving around the food on your plate. My experience is that a different process probably isn't the solution you're looking for.

4.3.4 Do the hard thing now

In this profession, we're often asked to deal with difficult situations. No set of rules can guide you safely through every scenario, but I have found that postponement is never the best solution.

Instead of avoiding the hardest parts, double down on them.

If you have a poor relationship with your manager or a member of your team, spend even more time with them. Meet with them every day, or have dinner with them. If two engineers are struggling to work together, before you separate them onto different teams get them to spend more time together trying to understand each other's perspective. (There are some obvious exceptions here, but if two people truly cannot work together, is there something else there that you've been avoiding dealing with?)

As a leader, you can't run from problems; engage 'em head-on.

4.3.5 Your company, your team, yourself

Lately, I've come to have something of a mantra for guiding decision-making: do the right thing for the company, the right thing for the team, and the right thing for yourself, in that order. This is pretty obvious on some levels, but I've found it to be a useful thinking aid.

First, all thinking should start from a company perspective, and you should make sure that what you're doing is not creating negative externalities for the company or the other teams you work with. For example, you're really excited about trying out a new programming language in a project, but you should also make sure that you've considered the additional maintenance cost for the rest of the company.

Next, make sure that your choices are being made on behalf of your team, not on your own behalf. This might mean pushing back on a timeline that would force your team into a death march, even though it's uncomfortable to have that conversation with your manager or your product partner.

Last in the list is yourself, but while I do believe that you should generally put yourself last, it's also a reminder to "pay yourself." Burnout is endemic in our industry, and a burned-out manager often leaks onto their team. Give as much as you can sustainably give, and draw the line there.

4.3.6 Think for yourself

So much of what we take for granted is cargo-culted instead of done with intention. Early in your management career, you'll have to figure out how to approach common challenges: interviewing, performance management, promotions, raises. It's totally all right to start out by following what you see around you—learning from your peers is critical to success. However, it's also important to be honest about which of your practices are truly best practices and which you're following on autopilot.

The recent focus on programming interviews[7] is a great example. Most hiring managers, myself certainly included, are aware that they're conducting mediocre interviews, but over time it's easy to lose that perspective. You can't fix everything at once, so you'll often be doing something mediocre at any given point in time, but remember to come back and improve it when you can (e.g., pay down your management debt).

As a final thought, the best management philosophy never stands still, but—in the model of the Hegelian dialectic[8]—continues to evolve as it comes in contact with reality. The worst theory of management is to not have one at all, but the second worst is one that doesn't change.

4.4 Managing in the growth plates

My last year at Digg, we were heads-down executing, trying to carve a path out of falling user numbers and an evaporating cash reserve. When we were acquired by SocialCode, I jumped right into execution mode, and instant conflict. What I'd learned over the past two years about leadership—execute, execute, execute—was disruptive, and I couldn't figure out why. I've seen the opposite happen just as often, when experienced, successful managers from well-established companies dive into a startup and exit soon thereafter with a legacy of ineffective initiatives.

The most confusing places to start are midsize, rapidly growing companies. That's because parts of the company are growing quickly, with an emphasis on execution, and other parts have largely stabilized, with ideas becoming the more valued currency. Long bones have growth plates at their ends, which is where the growth happens, and the middle doesn't grow. This is a pretty apt metaphor for rapidly growing companies, and a useful mental model when trying to understand why your behaviors might not be resonating in a new role.

4.4.1 In the growth plates

At a small startup or in a rapidly growing company's growth plates, you're mostly dealing with new problems. These new problems aren't necessarily novel (most problems are people problems), but they are problems that your company has never prioritized long enough to get a usable solution out the door. This means that you can't expect to succeed by iterating on the status quo.

You'd expect that novel ideas would be heavily valued in these circumstances, but, interestingly, it's the opposite: execution is the primary currency in the growth plates. That's because you typically have a surplus of fairly obvious ideas to try, and there is constrained bandwidth for evaluating those ideas.

It's common for well-meaning individuals from outside the growth plates to jump in to help by supplying more ideas, but that's counterproductive. What folks in the growth plates need is help reducing and executing the existing backlog of ideas, not adding more ideas that must be evaluated. Teams in these scenarios are missing the concrete resources necessary to execute, and supplying those resources is the only way to help. Giving more ideas feels helpful, but isn't.

Finally, I think it's important to recognize to recognize that, in the growth plates, you are focused on surviving to the next round, which might be a different growth challenge, or might be the team stabilizing. It is extremely hard to consistently do the basics well in these circumstances, because you simply won't have enough time to do them well. You'll have to get comfortable doing as well as time constraints allow, and sometimes that will lead to being mediocre at things you're passionate about. I personally find that I shift into working on the system,[9] and—embarrassingly and unfortunately—tend to cut down on many facets of people management.

4.4.2 Outside the growth plates

Away from the growth plates, you are mostly working on problems with known solutions. Known solutions are amenable to iterative improvement, so it would make sense for execution to be highly valued, but I find that, in practice, ideas—especially ideas that are new within your company—are most highly prized.

All slow-growth environments used to be high-growth environments, which means they were once run by someone who was a sufficiently effective executor to evolve them into a slow-growth environment. Consequently there tends to be less iterative improvement available than you'd expect. So often, we make solid executors responsible for slower-growth areas—we need the innovators in the highest-growth ones—but the opposite tends to work better.

As a manager, this is the environment for you to do the basics very, very well. Spend time building rich relationships, gelling your team, working with them on career development. Build up so that when innovation or external change pushes you off your local maxima, you and the team are ready and rested.

4.4.3 Aligning with values

The message I'd end with is a simple one: be thoughtful about carrying your values with you from one context into another. Leadership is matching appropriate action to your current context, and it's pretty uncommon that any two situations will flourish from the same behaviors. If you're working in the growth plates—or outside of them—for the first time, treat it like a brand-new role. It is!

4.5 Ways engineering managers get stuck

As a new manager, I found it useful to start each performance review season by rereading! Which also means it's an excellent time to reread Camille Fournier's "How Do Individual Contributors Get Stuck?"[10] Over time, I found myself wanting a manager-centric version, and eventually that desire solidified into the following list.

Following in Fournier's tradition, I was thinking about the parallels for engineering managers. Managers work more indirectly, so when we get stuck it isn't always quite as obvious, but we absolutely do get stuck, both on individual projects and in our careers.

Here are a few ways that happens.

Newer managers, often in their first couple of years:

1. **Only manage down.** This often manifests in building something your team wants to build, but which the company and your customers aren't interested in.

2. **Only manage up.** In Pearl S. Buck's *The Good Earth,*[11] she writes, "All power comes from the Earth." In management, power comes from a healthy team. Some managers focus so much on following their management's wishes that their team evaporates beneath them.

3. **Never manage up.** Your team's success and recognition depend significantly on your relationship with your management chain. It's common for excellent work to go unnoticed because it was never shared upward.

4. **Optimize locally.** Picking technologies that the company can't support, or building a product that puts you in competition with another team.

5. **Assume that hiring never solves any problems.** When you're behind, it can be tempting to spend all of your time firefighting and neglect hiring, but if your business is growing quickly, then eventually you hire or burn out.

6. **Don't spend time building relationships.** Your team's impact depends largely on doing something that other teams or customers want, and getting it shipped out the door. This is extraordinarily hard without a supportive social network within the company.

7. **Define their role too narrowly.** Effective managers tend to become the glue in their team, filling any gaps. Sometimes that means doing things you don't really want to do, in order to set a good example.

8. **Forget that their manager is a human being.** It's easy to get frustrated with your manager when they put you in bad situations, forget to tell you something important, or commit your team to something without consulting you, but they almost certainly did it with the best of intentions. To have a good relationship with your manager, you have to give them room to make mistakes.

More experienced managers:

1. **Do what worked at their previous company.** When you start a new job or new role, it's important to pause to listen and foster awareness

before you start "fixing" everything. Otherwise, you're fixing problems that may not exist, and doing so with tools that may not be appropriate.

2. **Spend too much time building relationships.** This is particularly common in managers coming from larger companies into smaller ones, and it creates the perception that the manager isn't contributing anything of value. This tends to be because smaller companies expect more execution focus than relationship management focus from their managers.

3. **Assume that more hiring can solve every problem.** Adding a few wonderful people to the team can solve many problems, but adding too many people can dilute your culture, and lead to people with unclear roles and responsibilities.

4. **Abscond rather than delegate.** Delegation is important, but it's easy to go too far and ignore the critical responsibilities that you've asked others to take on, only to discover an easily averted disaster later on.

5. **Become disconnected from ground truth.** Particularly at larger companies, it can become frequent to make decisions that appear to be fundamentally disconnected from reality.

Managers of any and all levels of experience:

1. **Mistake team size for impact.** Managing a larger team is not a better job, it's a *different* job. It also won't make you important or make you happier. It's hard to unlearn a fixation on team size, but if you can, it'll change your career for the better.

2. **Mistake title for impact.** Titles are arbitrary social constructs that only make sense in the context they're given. Titles don't translate across companies meaningfully, and they're a deeply flawed way to judge yourself or others. Don't let them become your goal.

3. **Confuse authority with truth.** Authority lets you get away with weak arguments and poor justifications, but it's a pretty expensive way to work with people, because they'll eventually turn off their minds and simply follow orders—if they're in a complicated compensation or life situation, that is. Otherwise, they'll just leave.

4. **Don't trust the team enough to delegate.** You can't scale your impact or engage your team if you don't give them enough room to do things differently than you would. Many organizations become bottlenecked on approvals, which is a sure proxy for lack of trust.

5. **Let other people manage their time.** Most managers have significantly more work they could be doing than they're able to do. This will probably be your status quo for the rest of your career, and it's important to prioritize your time on important things, and not simply on whatever happens to end up on your calendar.

6. **Only see the problems.** It can be easy to only see what's going wrong, and forget to celebrate the good stuff. Down this path lie frustration and madness.

I'm certain there are hundreds more ways that managers get stuck, but those are the ones that came to mind first!

4.6 Partnering with your manager

At my first software job, I chatted one-on-one with my manager twice in two years, including my first year, when I was remote and three time zones away. In that situation, you either become self-managing or you get let go for inactivity, and somehow I found things to do. (I'd like to add *useful* to *things*, but as best I can tell, my team's software was unilaterally thrown away, so that's hard to justify.)

Something that experience didn't equip me to do well is partner with my manager. I came away without a mental model for what management does, let alone how you would work with them. It's been a rocky path for me to figure out a healthier approach, and if you've faced a similar struggle, hopefully these ideas will help.

To partner successfully with your manager:

1. You need them to know a few things about you.

2. You need to know a few things about them.

3. You should occasionally update the things you know about each other.

Things your manager should know about you:

- What problems you're trying solve. How you're trying to solve each of them.
- That you're making progress. (Specifically, that you're not stuck.)
- What you prefer to work on. (So that they can staff you properly.)
- How busy you are. (So that they know if you can take on an opportunity that comes up.)
- What your professional goals and growth areas are. Where you are between bored and challenged.
- How you believe you're being measured. (A rubric, company values, some KPIs, etc.)

Some managers are easier to keep informed than others, and success hinges on finding the communication mechanism that works for them. The approach that I've found works well is:

1. Maintain a document with this information, which you keep updated and share with your manager. For some managers, this will be enough! Mission accomplished.
2. Sprinkle this information into your one-on-ones, focusing on information gaps (you're not seeing support around a growth area, you're too busy, or not busy enough, and so on). Success is filling in information gaps, not reciting a mantra.
3. At some regular point, maybe quarterly, write up a self-reflection that covers each of those aspects. (I've been experimenting with a "career narrative" format that is essentially a stack of quarterly self-reflections.) Share that with your manager, and maybe with your peers too!

A few managers seemingly just don't care, and I've always found that those managers do care, and are too stressed to participate in successful communication. This leads to the other key aspect of managing up: knowing some things about your manager and their needs.

Here are some good things to know:

- What are their current priorities? Particularly, what are their problems and key initiatives. When I get asked this question, I often can't answer it directly, because what I'm focused on is people-related, but it's a warning sign if your manager never answers it (either because because they don't know, or they are always working on people issues).

- How stressed are they? How busy are they? Do they feel like they have time to grow in their role or are they grinding?

- Is there anything you can do to help? This is particularly valuable for managers who don't have strong delegation instincts.

- What is their management's priority for them?

- What are they trying to improve on themselves, and what are their goals? This is particularly valuable to know if they appear stuck,[12] because you may be able to help unstick them. (You could be especially helpful by redefining impact in terms of work that your team can accomplish versus growing team size, which is a frequent source of stickiness!)

It's relatively uncommon for managers to be unwilling to answer these kinds of questions. (Either they're open and glad to share or are willing to speak about themselves.) However, it is fairly common for them to not know the answers. In those cases, each of these questions can be a pretty expansive topic for a one-on-one.

4.7 Finding managerial scope

I was chatting with an engineering manager last week, and he mentioned that the jobs he really wants are VP of Engineering roles, but he feels that no one is willing to take a gamble on him. Instead, he's looking at line management opportunities in fast-growing companies where he can go in with a small team and rapidly grow managerial scope as the company grows around him.

Rather than casting a stone, I promptly threw all my stones away: I've been guilty of pursuing exactly that strategy in a previous job hunt.

If you luck into a good situation, career progression can be so automatic early in your career that it can take a while to realize later in your career that your progression's slope has flattened out like a penny thrown off the Empire State Building and then run over by a coal train.

Broadly, there are three types of engineering management jobs:

1. *Manager*: you manage a team directly,
2. *Director*: you manage a team of managers,
3. *VP*: you manage an organization.

Especially early in your management career, it's easy to conflate reaching the next "rung" with reaching a certain number of people you've managed. Following this line of reasoning, for a 100-person company to hire you, you might need five direct reports to be a manager, 20 to be a director, and 40 to be a VP.

It's easier to focus on team size than title because the low cost of minting titles drives a fiercely inflationary economy. At Digg, I became a Director of Engineering because the company and my team kept shrinking. Far from a recognition of my success, this was a party favor for participating in one of history's great showcases of snatching defeat from the jaws of victory.

As managers looking to grow ourselves, we should really be pursuing scope: not enumerating people but taking responsibility for the success of increasingly important and complex facets of the organization and company. This is where advancing your career can veer away from a zero-sum competition to have the largest team and evolve into a virtuous cycle of empowering the organization and taking on more responsibility.

There is a lot less competition for hard work.

Companies will always need someone to run their cost-accounting initiatives, to set up their approach to on-call, to iterate on their engineer-recruiting process. Strong execution in these crosscutting projects will give you the same personal and career growth as managing a larger team. Project-managing an initiative working with 50 engineering managers is a far better learning opportunity than managing an organization of 50, and it builds the same skills.

This realization was very important and empowering for me: you can always find an opportunity to increase your scope and learning, even in a company that doesn't have room for more directors or vice presidents.

It also changed how I hire engineering managers, allowing me to switch from the pitch of managing a larger team as the company grows—an oversubscribed dream if there ever was one—to a more meaningful and more reliably attainable dream of growing scope through broad, complex projects.

If you've been focused on growing the size of your team as the gateway to career growth, call bullshit on all that,[13] and look for a gap in your organization or company that you can try to fill.

You'll be a lot happier.

4.8 Setting organizational direction

It took me two years as a manager to reach the "leadership is lonely" phase. Folks had warned me that it would happen, and it did. The team was struggling to acclimatize after acquisition, and I felt like I was carrying the stress alone. I saw the problems, but didn't know how to make progress on them. Two years later, I'd learned more about management, was increasingly able to rely on experience over invention, and was no longer lonely.

When I began managing managers, things shifted. I felt certain that I knew how to solve all the problems, but I didn't know how to rely on others to solve them, and often learned of problems long after they'd deteriorated. Delegation, metrics, meetings, and process—practices that I'd considered obvious or unimportant—crept into my tool kit, and I started to regain my footing.

Over the past year, as I've transitioned into largely working with managers of managers, things have shifted again. Let's explore that a bit.

4.8.1 Scarce feedback, vague direction

For much of your early career, you'll have folks who are routinely giving feedback on your work. As your span of responsibility grows, particularly if it's somewhat specialized, increasingly no one will feel responsible for or able to provide that feedback. In a new function, at a small company, your team might be two people, and already you're inhabiting this realm of muted feedback.

Where you used to get direct, actionable advice, now you're listening for ghosts: grumbles on your team about too much technical debt, the rumor that two peers are engaged in some light feuding, agitation where there was previously calm.

As a functional leader, you'll be expected to set your own direction with little direction from others. When things in your area are going poorly, you'll be swamped with more direction and input than you can readily absorb, but when things are going well, you'll often be responsible for supplying your own direction and that of your team.

If you don't supply it yourself, you'll start to feel the pull of irrelevance: Maybe no one really cares what we do? What would happen if I stopped showing up? Maybe I should be doing something different?

That initial instinct to leave after hitting a pocket of seeming irrelevance is a comforting one, but it's the wrong way to go. You can certainly avoid the current swells of ambivalence by switching jobs, but if you're successful at another company then you'll end up in the same situation.

This is a symptom of success. You have to learn the lesson it's trying to teach you: How to set your organization's direction and your own.

4.8.2 Mining for direction

The first step to setting direction is to cast the widest possible net for ideas. Talk to folks at your company who have worked at different kinds of companies, and ask what those companies did really well. Talk to your team and see if there are ideas to draw out that they've been noodling on but haven't yet volunteered. Read some new technical papers.[14] Meet with peer companies and ask them what they're focused on. Do the same with the Googles/Facebooks and the smallest, most interesting companies.

This first phase is discovery without judgement. You should take ideas from everywhere and generate a pile of ideas that folks are pursuing, even if you think they're terrible.

Once your pile of ideas has gotten large enough, craft it into a strategy,[15] and then start testing that strategy. Keep refining and exploring your strategy until you can figure out what the key decisions are, a kind of ad hoc sensitivity analysis.[16] Once you identify the key pivots in your strategy, you're finally prepared to define a direction!

Make a clear decision on each of those pivots, write up a document explaining those decisions, and then see if you can get anyone to read it. They'll disagree with a lot of what you've written, or else they'll be confused by it. Keep testing, and refine the confusion down to the smallest group of controversial problems possible.

Once you have those problems, return to your rounds of engaging with experienced leaders at other companies, and ask them how they've made those trade-offs before. Ask them for their stories. Ask them for the context that made one path perfect early on, and why they changed their minds as their company grew larger.

Incorporate everything you've learned into your strategy document, and you're done.

Well, almost. The one remaining problem is that almost no one will have the time to soak up the full detail of your overly precise document, so the final step is to distill it into something comprehensible without hours of close reading. I'm still working on the best way to do this, but I suspect that it's cutting all unnecessary, and even most essential, complexity in order to capture as much of the meaning as possible in three to four bullet points.

4.9 Close out, solve, or delegate

In a recent chat with an engineering manager, they mentioned a low-grade, ambient anxiety around their impact, which they'd felt since moving into a role focused on supporting managers rather than on leading a team directly. I think that this resonates with everyone's transition to managing managers: it's an unsettling period when you lose access to what used to make you happy—partnering directly with a team—and haven't found new sources of self-worth in your work.

This isn't the only reason this transition is hard, it's also hard because a lot of your skills and habits stop working well. The skill that scales the worst is outworking your problems.

This is particularly frustrating, because your ability to put your head down and solve gritty, important problems was probably a big part of how you were promoted. Now it's the wrong answer to most of your problems. This isn't because it's bad behavior, just because it's too inefficient to accomplish the breadth and quantity of things you need to get done.

If you're sitting at the post-transition moment, detached from the work you loved, and with your instincts driving you into a pile of work you can't make a dent in, I have a tool that's been useful for me and might be useful for you!

For every problem that comes your way—an email asking for a decision, a production problem, a dispute around on-call, a request to transfer from one team to another—you must pick one of three options:

Close out. Close it out in a way such that this specific ask is entirely resolved. This means making a decision and communicating it to all involved participants. This strategy is a success if this particular task never comes back to you; and your goal is to finish this particular task as quickly and as permanently as possible.

Solve. Design a solution such that you won't need to spend time on this particular kind of ask again in the next six months. This is often designing norms or process, but depending on the kind of issue, this might be coaching an individual. With this option, your goal is to finish off an entire category of tasks.

Delegate. Ideally, this is to redirect the ask to someone who is responsible for that kind of work, but sometimes it is a one-off. If you can't close out a task or solve it, your only other option is to delegate it to someone who either has the specialized skills to close/solve it, or who can work in the system.

No matter what problem comes your way, you're not allowed to solve issues any other way! Give this method a try for a week, and see if it helps you navigate your role more effectively.

5

CHAPTER 5

Culture

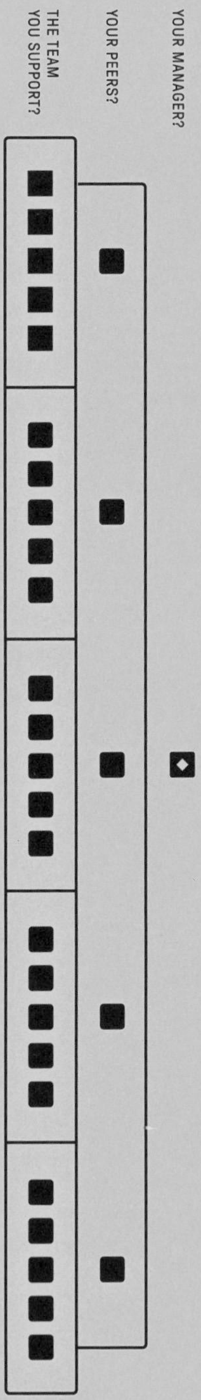

Figure 5.1
Membership in multiple teams: peers, managers, and the supported team.

Culture

When I'm working on a presentation to a large organization, I spend a great deal of time on framing it properly. There are so many perspectives that I want my message to resonate with, and I want folks to walk away humming the right notes. However, in the long run our measure is not in what we say or how skillfully we say it, but in what we do, and the abacus tallying our actions is organizational culture.

Sometimes that tally doesn't show a result that we're proud of, but the good news is that culture evolves. Nonetheless, this evolution only yields positive change to persistent effort. In this chapter, we explore some of the persistent efforts I've used to shift culture in organizations that I've supported.

5.1 Opportunity and membership

For a long time, I've found the idea of fostering an inclusion organization to be somewhat intimidating. For most tasks, I've been able to plan out a roadmap, identify some reasonable metrics, and get to work, but for inclusion I was apt to stare at the blank page, filled with uncertainty.

Since then, I've found a framework for thinking about inclusion efforts that is simple but that has allowed me to think about the problem broadly, identify useful programs, and move from anxiety to implementation. The basis: an inclusive organization is one in which individuals have access to opportunity and membership. Opportunity is having access to professional success and development. Membership is participating as a version of themselves that they feel comfortable with.

I've found this framework powerful for reflecting on what is going well and where we can improve, and I hope you'll find it useful as well. For both themes, I've written notes on investment and measurement.

5.1.1 Opportunity

There are workplaces where everyone around you is delightful, the customers are friendly, and you feel respected, but you still return home each night dissatisfied. Occasionally an interesting project will come up, but those typically go to more tenured folks. When I think about having access to opportunity, I think about ensuring that folks can go home most days feeling fulfilled by challenge and growth.

The most effective way to provide opportunity to the members of your organization is through the structured application of good process. Good process is as lightweight as possible, while being rigorous enough to consistently work. Structure application allows folks to learn how the processes work, and lets them build trust by watching the consistent, repeated application of those processes.

As the saying goes, this is simple to do but far from easy. The key question is whether you'll continue to respect your processes when it's inconvenient to do so. If one of your best team members wants a specific opportunity, are you willing to run an open application process? What if they plan to leave your company otherwise? Would you be willing to bypass your processes to keep them?

There are infinite ways to create and distribute opportunity! Some of the programs that I have found more helpful are:

1. **Rubrics everywhere.** Every important people decision should have a rubric around how folks are evaluated. This is true for promotions, performance designations, hiring, transitions into management, and pretty much everything else!

2. **Selecting project leaders.**[1] Having a structured approach to selecting project leads allows you to learn from previous selections, and to ensure that you're not concentrating opportunity on a small set of individuals.

3. **Explicit budgets.** Many companies take a "spend it like your own money" approach to budgets, which often leads to large inconsistencies across individuals. Instead of saying that you'll pay for teams to attend a reasonable number of conferences per year, specify a fixed number. Instead of maintaining a general ongoing education budget, make it explicit.

4. **Nudge involvement.** Many people are uncomfortable applying to opportunities, using education budgets, asking for mentorship, etc. It's very effective to reach out to those folks directly and recommend they apply. Even more powerful is showing them where they fall on a distribution: it's one thing to know that you've never used your educa-

tion budget, and something else entirely to know that you're the only person who isn't using it.

5. **Education programs.** Create ongoing training and learning programs that are available for everyone, or, for example, for all managers.

   I'm pretty confident that these measures will significantly improve the distribution of and access to opportunity, but we can do better. We can measure. I've found measurability of opportunity to be surprisingly high, which is one of the reasons I think it's an effective pillar in thinking about inclusion.

   The metrics that have been useful for me:

1. **Retention** is the most important measure of availability of opportunity, although it's also a very lagging indicator. This should be the first thing you're paying attention to, but you must recognize that it's slow to show change.

2. **Usage rate** is how often folks get picked in project selection.[2] The number of distinct team members picked to lead critical projects is a particularly interesting measure.

3. **Level distribution** is useful, in particular comparing cohorts of folks with different backgrounds. People want role models for themselves in senior roles at the company where they work, which is why looking at the representation of underrepresented minorities and women is only a start. You also want to look in each role and level of seniority.

4. **Time at level** is how long team members wait between promotions. How does this compare across cohorts? Something to watch for is that this number is also greatly influenced by initial level at hiring. For example, time at level might look equivalent because some cohorts are being under-leveled at time of hire.

   Getting access to some of this data will require partnering with your human resources team, but I've found that friendly persistence, along with sharing your thinking behind the asks, works well to get coworkers working with you.

5.1.2 Membership

Membership is a bit harder to measure, but equally important. I can remember once having coffee with a coworker who described their daily calculus of trying to find someone to eat lunch with. Their work was generally going well, but each day, as noon approached, what they thought about most was feeling lonely.

If you're spending so much energy wondering whom you'll eat lunch with, that's energy you can't spend being creative. If the idea of going to work gives you anxiety, at some point you're going to decide to stop coming. Membership is the precondition to belonging.

The programs I've found most impactful here:

1. **Recurring weekly events** allow coworkers to interact socially. These are held during working hours, are open to folks from many different teams to attend, and are optional. One of my personal favorites is hosting a paper-reading group.[3]

2. **Employee Resource Groups (ERGs)** create opportunities for folks with similar backgrounds to build community.

3. **Team offsites** once a quarter or so are a good chance to pause, reflect, and work together differently. Spending a day together learning and discussing is surprisingly effective at making individuals feel like a team. This is particularly true for groups of managers, whose daily cadence is typically more in tune with their teams than with their peers.

4. **Coffee chats**, perhaps randomly assigned by Donut,[4] get folks from different teams interacting when they don't need something from each other.

5. **Team lunches** give folks time to relax a bit and interact socially. Held once a week or so, they can become a pleasant ritual. These can be a bit risky, at least for folks (like early hires) who feel uncomfortable in their team. Navigating the social mores of an entire team can be much more difficult than in a one-on-one.

These programs are all simple, and their simplicity can hide the degree of thoughtful care necessary to do them well. Depending on where a team or organization is, you'll have to adjust your approach to make them effective. Always take the extra day to test your implementation proposal with a variety of coworkers.

As you roll these out, measuring remains extremely important, although I've found membership rather harder to measure than opportunity. Here are some of the potential measures:

1. **Retention** is once again the golden measure, and once again a long-trailing indicator.

2. **Referral rate** by cohort provides insight into which individuals feel comfortable asking their friends and previous coworkers to join the company.

3. **Attendance rates** for recurring events and team lunches provide some insight into whether folks feel comfortable with those groups.

4. **The quantity and completion rate of coffee chats** are automatically measured with Donut.

Just as when you collect the data to measure opportunity, this will require some partnership with human resources, but it's well worth the effort.

A second similarity between the two is that balancing opportunities for membership across a large population is pretty tricky. Many activities and events don't work well for everyone—meals can be difficult for individuals with complex dietary restrictions, physical activities make some uncomfortable, activities after working hours can exclude parents—and success here requires both a broad portfolio of options and a willingness to balance concerns across events and time.

5.1.3 Keep going

Combine efforts on opportunity and membership, and you will find yourself solidly on the path to an inclusive organization. This strategy doesn't have much flash, but results are louder than proclamations. The most important thing is continuing your investment over the long term. Pick a few things that you are able to sustainably continue, get started, and layer in more as you build steam.

5.2 Select project leads

Have you ever worked at a company where the same two people always got the most important projects? Me too. It's frustrating to watch these opportunities to learn from the sidelines, and reliance on a small group can easily limit a company's throughput as it grows. This is so important that I've come to believe that having a wide cohort of coworkers who lead critical projects is one of the most important signifiers of good organizational health.

It's a particularly powerful metric because it simultaneously measures the company's capability to execute projects and the extent that its members have access to growth. The former measurement helps determine the company's potential throughput, and the latter correlates heavily with inclusivity.

In this context, there are two kinds of projects: critical projects and everything else. Critical projects are scarce. There are more people who want them than can get access to them. Other projects are abundant. You might not be able to get one immediately, but if you wait a month or two the odds are good. There's no need to be structured about abundant projects!

To increase the number of folks leading this kind of project, I've iterated into a structured process that has worked quite well:

1. **Define** the project's scope and goals in a short document. Particularly important are identifying:

    - *Time commitment.* People need to decide if they must ask permission from their managers.

    - *Requirements to apply.* If there are no requirements, say so explicitly; otherwise, a lot of folks will assume that there are, and will opt out.

    - *Selection criteria.* If multiple individuals apply, how will you select the project leader between them?

2. **Announce** the project to a public email list, at an all-hands, over Slack, or however your company does persistent communication; I tend to use email for these. What's most important is that you:

- Allow folks to *apply in private.* Some individuals will be uncomfortable applying in public.
- Make sure that applicants *don't see who else has applied.* Some people will see someone they view as senior apply, and will immediately disengage because they feel that they are less qualified.
- Give *at least three working days for people to apply.* Do this whenever possible, as some folks will need to talk with their manager or peers to muster up the confidence to apply.

3. **Nudge** folks to apply who you think would be good candidates but who might not self-select. This is particularly important for getting new people into the process.
4. **Select** a project leader based on the selection criteria you specified. Take the time to consider every single applicant against the criteria, and, if possible write up a paragraph or two about each of them. Once you've selected the leader, privately reach out to them to confirm that they're able to commit.
5. **Sponsor** the project leader by finding someone who has successfully completed a similar project to be their advisor. This sponsor will be accountable for coaching the leader to successful completion. This is a great learning opportunity for sponsors, as they are typically folks who are great at doing things themselves, but not as used to teaching others how to lead large and ambiguous projects.
6. **Notify** other individuals who applied that they were not selected. It's extremely helpful if you provide them feedback on why you didn't select them. Sometimes it's because they've already done something great and you want to create room for another person to learn, and that's a totally reasonable thing to tell them!
7. **Kick off** the project, notifying the same folks who received the application announcement who the project leader is, who the sponsor is, and what their plan for running the project is.
8. **Record** the project, who was selected, and who the sponsor is into a public spreadsheet. Also link out to a project brief!

If you do this, over time you'll get a clear sense of who gets leaned on for the most important projects. If you do this well, you'll see that cohort continue to grow!

The first few times you do this, it will feel very constraining and inefficient. Previously, you would have just sent a ping to a favored individual and they'd have been off and running, but now you have to run a slower and more deliberate process. Increasingly, though, I believe this is the most important change in my approach to leadership over the past few years. Done well, it can be the cornerstone in your efforts to grow an inclusive organization.

5.3 Make your peers your first team

While companies are literally composed of teams,[5] I've found it surprisingly common to meet folks who feel as if they are not a member of any team. Managers working directly with engineers tend to feel some membership in that team, but even the rarest occasions of authority create a certain distance. Managing managers lets you pierce the veil a bit more, but the curtain falls when it comes to performance management or assigning resources.

Looking toward your peers can be uncomfortable in a different way. As your span of responsibility grows, you may not know much of your peers' work, or you may find yourself frequently contesting against them for constrained resources. Even when surrounded by the fastest of growth, you may be awkwardly aware that you're aspiring to move into the same, rare, roles.

These dynamics can lead to teams whose camaraderie is at best a qualified non-aggression pact, and in which collaboration is infrequent. It's a strange tragedy that we hold ourselves accountable for building healthy, functional teams, and yet are so rarely on them ourselves.

But it can be better. It's okay to expect more.

I've worked on a few teams in which folks consistently looked out for each other, and believed they'd all come out better together. These were teams that had individuals willing to disappoint the teams they managed in order to help their peers succeed. It wasn't that these folks were ready to callously disappoint people. Rather, they balanced outcomes from a broad perspective that included their peers.

The members of such a team have shifted their *first team* from the folks they support to their peers. While these teams are rare, I've come to believe that supporting their creation is simple—albeit hard—work, and when the conditions are fostered, they lead to an exceptionally rewarding work environment.

The ingredients necessary for such a team are:

Awareness of each other's work. Even with the best intentions, a member cannot optimize for their team if they're not familiar with other members' work. The first step to moving someone's identity to their peers is to ensure that they know about their peers' work. This will require a significant investment of time, likely in the form of sharing weekly progress, and the occasional opportunity for folks to dive deep into each other's work.

Evolution from character to person. When we don't know someone well, we tend to project intentions onto them, casting them as a character in a play they themselves are unaware exists. It's quite challenging to optimize on behalf of characters in your mental play, but it's much easier to be understanding of people you know personally. Spending time together learning about each other, often at a team offsite, will slowly transform strangers into people.

Refereeing defection. In game theory, there is an interesting concept of a *dominant* strategy. A *dominant* strategy is one that is expected to return the maximum value regardless of the actions of other players. Team collaboration is not a dominant strategy. Rather, it depends on everyone participating together in good faith. If you see someone acting against the interests of the team, you, too, will likely defect to pursue your own self-interest. Some teams are tightly knit enough that no one ever attempts defection, but most teams experience frequent changes in membership and external conditions. I believe that on such teams coordination is only possible when the manager or a highly respected member operates as a referee, holding team members accountable for good behavior.

Avoiding zero-sum culture. Some companies foster zero-sum cultures, in which perceived success depends on capturing scarce, metered resources, like head count. It's hard to convince folks to coordinate under such conditions. Positive cultures center on recognizing impact, support, and development, which are all avenues that support widespread success.

Making it explicit. If you have the first four ingredients, you still have to explicitly discuss the idea of shifting folks' identities away from the team they support and toward the team of their peers. It's hard to move membership from the team you spend the most time with, and I haven't seen it happen organically.

Given how much energy it takes to shift the identities of a team of managers away from the folks they support and toward their peers, I think it's quite reasonable to question whether it's genuinely worth doing it. You'll be unsurprised to know that I think it is.

As you move into larger roles, you'll need to start considering challenges from the perspectives of more teams and people. In this sense, treating your peers as your first team *allows you to begin practicing your manager's job*, without having to get promoted into the role first. The more fully you embrace optimizing for your collective peers, the closer your priorities will mirror your manager's. Beyond practicing working from this broader lens, it will also position you for particularly useful feedback from your manager, as you'll both be considering similar problems with shared goals.

The best learning doesn't always come directly from your manager, and one of the most important things a first team does is provide a *community of learning*. Your peers can only provide excellent feedback if they're aware of your work and are thinking about your work similarly to how you are. Likewise, as you're thinking about your peers' work, you'll be able to learn from how they approach it differently than you anticipate. Soon, your team's rate of learning will be the sum of everyone's challenges, no longer restricted to just your own experiences.

Long term, I believe that your career will be largely defined by getting lucky and the rate at which you learn. I have no advice about luck, but to speed up learning I have two suggestions: join a rapidly expanding company, and make your peers your first team.

Jason Wong's "Building a First Team Mindset"[6] is an excellent read on this theme if you're looking for more!

5.4 Consider the team you have for senior positions

Over the past six months, I've been hiring engineering manager-of-managers roles. These roles are scarcer than line management roles, and they vary more across companies. This process has taught me a bunch of new things.

Manager-of-managers searches are interesting in at least four ways:

- There are many folks who can't find upward mobility within their current company. They have not managed managers before, and are looking for the opportunity.

- Most people with experience managing managers are happy in their current roles.

- The individuals interested in these roles outpace supply. This makes it more important to put in place processes like the Rooney Rule.[7]

- You need a fair way to consider candidates within your company. It must be respectful to them yet allow you to uphold your responsibilities to the company.

The last aspect what has taught me the most, and what I want to focus on. Ensuring that internal candidates take part is essential to an inclusive culture. Fair consideration doesn't mean that we prefer internal candidates. Rather, it means that there is a structured way for them to apply, and for us to consider them.

Letting individuals apply is the easy part. You must announce each position and ask for internal candidates. You should persuade eligible candidates to apply, especially if they are uncertain. You should give them a week or two to consider whether they want to apply.

Then comes the trickier part: evaluation. We've focused on testing these categories:

1. **Partnership.** Have they been effective partners to their peers, and to the team that they've managed?

2. **Execution.** Can they support the team in operational excellence?

3. **Vision.** Can they present a compelling, energizing vision of the future state of their team and its scope?

4. **Strategy.** Can they identify the necessary steps to transform the present into their vision?

5. **Spoken and written communication.** Can they convey complicated topics in both written and verbal communication? Can they do all this while being engaging and tuning the level of detail to their audience?

6. **Stakeholder management.** Can they make others, especially executives, feel heard? Can they make these stakeholders feel confident that they'll address any concerns?

This evaluation doesn't cover *every* aspect of being an effective senior leader. But it does cover the raw skills that form the foundation of one's success. You already know if an internal candidate has hired managers. You know if they've done organization design. There's no need to ask all that.

To test these categories, we're using these tools:

Peer and team feedback. Collect written feedback from four or five coworkers. Include peers on other teams. Include people the applicants have managed. Include people they would not have managed. My biggest advice? Lean into controversial feedback, not away from it. Listen to would-be dissenters, and hear their concerns.

A 90-day plan. The applicant writes a 90-day plan of how they'd transition into the role, and what they would focus on. They emphasize specific tactics, time management, and where they'd put their attention. This is also a great opportunity to understand their diagnosis of the current situation. Provide written feedback to them on their plan. Have them incorporate that feedback into their plan. This is an opportunity to try out working together in the new role.

Vision/strategy document. The applicant writes a combined vision/strategy document. It outlines where the new team will be in two to three years, and how they'll steer the team to get there. Provide written feedback on the document. Have them incorporate that feedback.

Vision/strategy presentation. Have the applicant present their vision/strategy document to a group of three to four peers. Have the peers ask questions, and see how the applicant responds to this feedback.

Executive presentation. Have the applicant present their strategy document, one-on-one, with an executive. In particular, test for their ability to adapt communication to different stakeholders.

Running the process takes a lot of time, but it's rewarding time. In fact, this has generated more useful feedback than anything else I've done over the past year. It brings an element of intentional practice that's uncommon in engineering management. Folks get to take risks in their plans. You get to give direct feedback without risk of micromanagement. It's been useful enough that I'm now figuring out whether we can use a similar format for training managers.

Know that internal processes are awkward. You'll have many internal candidates. They will talk to each other. They will interview external candidates for a role they are applying for themselves. One candidate may end up managing the others. Don't decide to avoid this morass of awkwardness. You can't. You're deferring it until later. It'll still be awkward, but now awkwardly in the realm of gossip.

Running this process, and enduring the awkwardness in doing so, is the most rewarding thing I've done as a manager. I recommend it.

5.5 Company culture and managing freedoms

In 1969, Roger Miller (and later, much more famously, Janis Joplin) sang:

"Freedom's just another word for nothing left to lose." [8]

This gives us yet another peculiar entrant toward defining freedom. Perhaps it's nothing more than an apologetic[9] ode to despair: law requires consequences, and there are no consequences when you've already been reduced to nothing.

Lousy hooks and confusing segues aside, the topic for discussion is the relationship between a company's culture and its freedoms. Rather than starting down the path of examining the kinship of freedom and consequence—which is a rather dismal place to begin anything—we instead look toward kinds of freedom, which immediately brings us to the distinction between positive and negative freedoms.[10]

Positive freedom is your freedom to *do*: to vote, to wear the clothing you want, to own arms, to blow smoke into your neighbor's porch when they're trying to read a book outside on a sunny day. Negative freedom is your freedom *from* things: from being forced to take an impossible literacy exam before being allowed to vote, from to wear clothing you dislike or find oppressive, from having your cellular traffic recorded, from having your neighbors blow smoke onto your porch when you're trying to read a book outside on a sunny day.

Wielding this distinction, "freedom" is neither inherently good nor inherently just, and descends into the murky gray that already embroils everything else in our lives. Each positive freedom we enforce strips away a negative freedom, and each negative freedom we guarantee eliminates a corresponding positive freedom. This sad state of affairs is often referred to as the Paradox of Positive Liberty.[11]

I believe that the balancing of positive and negative freedoms is a fundamental task of managers and management. When we've lucked upon (or perhaps nurtured, if you're much more talented than I) a phenomenal culture and a great team who are executing well along a worthy roadmap, then, like a central bank reducing interest rates to avoid a bubble, or like a jogger reducing their pace to lower their heart rate, we can carefully ramp toward negative freedom and away from positive freedom. This is one of our essential tools for facilitating and prolonging success.

Further down the road, if the structure loses its luster, the economy shifts around us, or entropy's endless march throws a wrench into the machinery, then once again we shift toward positive freedoms, which gives the organization a greater chance to successfully adapt to its new circumstances.

Using the two together, management slowly decelerates to keep the good times rolling, and accelerates to help push through challenging periods.

Freedom is a loaded term, so it's easy to deteriorate into a moral discussion, but in times and topics of great sensitivity, I believe looking through the lens of system dynamics[12] is a valuable approach. Companies are vastly complicated systems with dozens of feedback loops, and managing the kind and quality of freedom is simply another mechanism to be adjusted, albeit with immense care and consideration.

A few closing tangents. First, Tom DeMarco's *Slack*[13] has an excellent suggestion for a good starting state between positive and negative freedoms for engineering teams: generally follow the standard operating procedure (i.e., keep doing what you're already doing, the way you're doing it), but always change exactly one thing for each new project. Perhaps use a new database, a new web server, a different templating language, a static JavaScript front-end, whatever—but always change exactly one thing.

Second, I'm always terrified of getting on the wrong side of history, so I've spent some time considering how this discussion of freedoms relates to Ben Horowitz's recent post on "Can Do vs. Can't Do Cultures."[14] I read that article as describing how young companies that are focused on innovation differ from mature companies that are trapped in "the innovator's dilemma."[15] Older companies can (and do) foster sheltered pockets of innovation, as in the example of Larry Page investing in good ideas that he encounters within Google, but maintaining a market position is fundamentally distinct from creating new markets. I think that the more complete argument is to use both cultures (and, in parallel, to place emphasis on positive and negative freedoms) in the appropriate circumstances.

5.6 Kill your heroes, stop doing it harder

The project launch is 18 months late, company revenue is dropping significantly, key people are leaving and being replaced by new hires. What to do? Well, work harder!

Does it work?

Of course not. Unless your problem is that people aren't trying hard, the "work harder" mantra only breeds hero programmers whose heroic ways make it difficult for nonheroes to contribute meaningfully. Later, as your new heroes finish martyring themselves on burnout, you're left with three exceptionally challenging problems:

1. You've bred a cadre of dissatisfied and burned-out heroes.

2. You and your heroes have alienated everyone else.

3. Your project is still totally screwed.

   This is a recurring pattern that many growing companies fall into, and it also happens to projects within larger companies. Anywhere you find managerial desperation and a hardworking team, "Do It Harder" may be visiting.

5.6.1 The fall and rise of a hero

One rainy day, you walk into the office and your boss wants to talk to you. He needs you to finish your project, but he also wants you to finish your coworker's project, without the coworker feeling bad about it, because your coworker is still going to *own* the project, you're just going to *do* it (along with your other work, remember).

A few weeks later, the site starts crashing every few days, and the company really needs to launch the new version of the site. The boss pops in to let you know he deeply trusts you, and needs you to take over both efforts. You're a good guy and it sounds like a good opportunity, and you're pretty sure that you can do a better job than the guys already working on it, so you say yes.

Congratulations! You're a hero programmer.

You're now working on five disparate projects, trying not to piss too many people off, but you're having trouble getting everyone involved. It seems like they're not really working as hard as you are, and it's a bit of a drag, since you're pulling 70-hour weeks and getting paged every Saturday night.

The other developers are glad that you're taking a lead on the problems that were terrifying them too, but all is not well. A couple are quietly bitter because of your newly elevated status, but most just don't know how to contribute anymore, because you and your hero peers are rewriting the existing system, debugging the outages, and cherry-picking the easy wins. What's left for them to do?

Day after day and week after week, the frustration between heroes and nonheroes grows stronger, tumbling toward inevitable disaster.

5.6.2 Kill the hero programmer

When it comes to solving the problem of the hero programmer, your options are limited: either kill the environment that breeds hero programmers, or kill the hero programmer by burnout.

They're truly unmaintainable beings, as their presence limits the effectiveness of those around them in exchange for a short-term burst of productivity fueled by long hours and minimized communication costs (minimized because most other people aren't able to do much).

These long hours burn your heroes out, and then they either quit or you shove them into a corner, where they'll glower at you while remembering how their hard work and critical contributions culminated in them glaring at you from that corner.

You can rehabilitate heroes, but it's touch-and-go from the beginning, and healing takes time. They might hate you for a while, and they probably should, because you created them with your ham-fisted attempt to fix your current problem.

5.6.3 A long time coming, a long time going

One of the observations from systems thinking[16] is that, though humans are prone to interpreting events as causal, often problems are better described in terms of a series of stockpiles that grow and shrink based on incoming and outgoing flows. The Dust Bowl[17] wasn't caused by one farmer or one year of overfarming, but by years of systemic abuse.

Stocks and flows are especially valuable in understanding the failure of projects and teams. Projects fall behind one sprint at a time. Technical debt strangles projects over months.

Projects fail slowly—and fixing them takes time, too.

Working at a frenetic pace for a couple of weeks or a month may feel like a major outpouring of effort and energy, but it's near impossible to quickly counteract a deficit created over months of poor implementation or management choices.

If hard work and breeding heroes doesn't work, what does?

5.6.4 Resetting broken systems

Your options for addressing a broken system depend on whether you're in a position to set policy. If you set the original direction and have the leverage to change directions, then resetting is as simple as standing up and taking the bullet for the fiasco you're embroiled in.

Taking the blame is painful, and it only plays well with the crowds a couple of times. After that, people won't trust you to lead them toward success, which makes some sense, since at that point you've led them off the rails multiple times. Fair's fair.

If taking the loss doesn't leave you looking shiny, at least attaining closure is healing, and the team will have the opportunity to start healing as the schedule is reset and goals are adjusted. Without the leverage to change policy—and this doesn't have to be direct authority, for influence is a powerful thing—you can't start the healing, but you can help reach the reset point more quickly. (This is similar to the protagonists' goal in Isaac Asimov's *Foundation's* series, as they struggle to accelerate and minimize the collapse of the Galactic Empire.[18])

Without policy, your tool is stepping back and allowing the brokenness to collapse under its own weight. A deeply flawed system can't be saved by band-aids, but it can easily absorb your happiness to slightly extend its viability. If you step back, you conserve your energy and avoid creating rifts by pushing others away in hero mode, and you will be ready to be a part of a new—hopefully more functional—system after the reset does occur.

This is a very uncomfortable process, and if you're a hardworking, loyal person, then it probably goes deeply against your nature. It certainly goes against my nature, but I believe that this is one case in which following my nature is a detriment to both myself and those around me.

Projects fail all the time, people screw up all the time. Usually it's by failing to acknowledge missteps that we exacerbate them. If we acknowledge errors quickly, and cut our losses on bad decisions before burning ourselves out, then we can learn from our mistakes and improve.

Kill your heroes and stop doing it harder. Don't trap yourself in your mistakes, learn from them and move forward.

6

CHAPTER 6

Careers

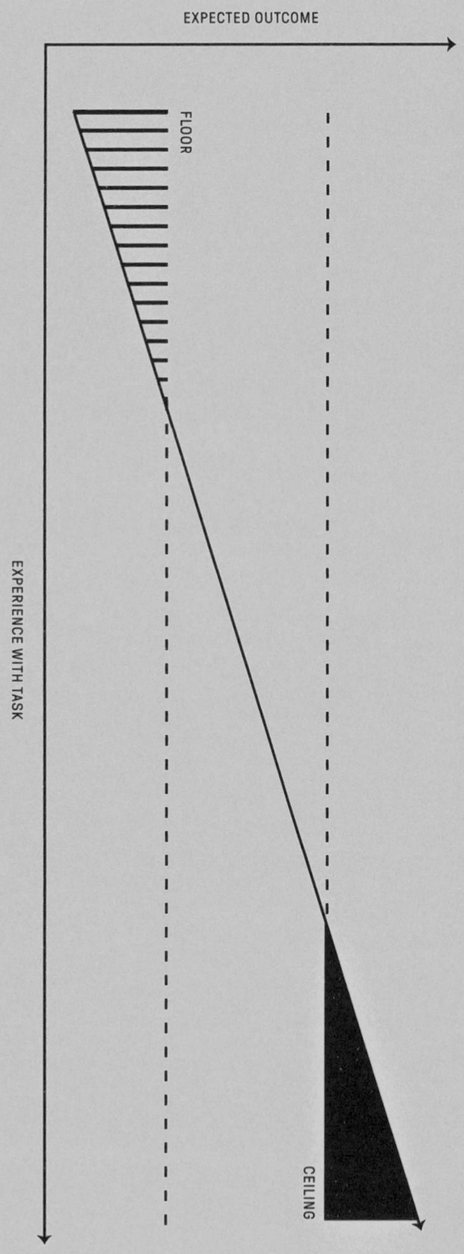

Figure 6.1
Relationship of practitioner experience with policies that raise floor or raise ceiling of expected outcomes.

Careers

Luck plays such an extraordinary role in each individual's career progression that sometimes the entire concept of career planning seems dubious. However, as managers, we have an outsize influence in reducing the role of luck in the careers of others. That potential to influence calls us to hold ourselves accountable for creating fair and effective processes for interviewing, promoting, and growing the folks we work with.

This chapter explores how we can design effective interviewing and hiring processes, as well as how we can steer our own careers while standing on a bedrock of constant change.

6.1 Roles over rocket ships, and why hypergrowth is a weak predictor of personal growth

There's a pervasive trope that people who've worked for an extended period in a large company will struggle to adapt to working in a smaller company. Work at a company too long, the theory goes, and you're too specialized to hire elsewhere. This belief is reinforced by both age bias[1] and the reality that few companies continue to win the rounds of reinvention necessary to maintain excellence over time. The pool of once-phenomenal companies is quite large: Yahoo!, Oracle, and VMware, to name a few.

If long tenure holds a stigma, what of short tenure? Yep, that's stigmatized, too.[2] Although, it's certainly less stigmatized than it used to be. A fairly consistent belief across companies is that multiple stints below one year are a bad sign, but, generally as long as you've worked a few years somewhere, then having a career that is otherwise entirely composed of one-year stints raises few red flags. While I'm certain that these beliefs are, at best, deeply flawed, they've accidentally become a rule of thumb in my own career: stay everywhere at least two years, and look for a place where I could spend four years to serve as the counterweight to my series of two-year stints. I've followed this rule very literally,[3] staying at least two years (well, we can all count here, *exactly* two years) at each company I've worked at, and starting each job with the hope that this one would be the place I'd make it to four.

To the surprise of no one, it turns out that's been a pretty terrible way to think about career planning, and lately I've been trying to find a more useful framework.

Working at a company isn't a single continuous experience. Rather it's a mix of stable eras and periods of rapid change that bridge between eras. Thriving requires both finding a way to succeed in each new era and successfully navigating the transitional periods. You yourself trigger some transitions, like switching companies. Others happen on their own schedule: a treasured coworker leaves, your manager moves on, or the company runs out of funding.

Discard the discussion of tenure. Let's talk about eras and transitions.

6.1.1 Your new career narrative

Start by building out a map of your past year. Each time there was a change that meaningfully changed how you work, mark that down as a transition. These could be your direct manager changing, your team's mission being redefined, a major reorganization, whatever—what counts is if how you work changed. What skills did you rely on to navigate the transition? What skills did the transition give you an opportunity to develop?

Next, think about the eras that followed those transitions. How did the values and expectations change? Did operational toil become considered critical work? Did work around inclusion and diversity become first-shift work that got mentioned in performance reviews? What skills did you depend on the most, and which of your existing skills fell out of use?

What you just wrote down is your new career narrative, and it's much richer than just another year at your company.

6.1.2 Opportunities for growth

The good news is that **both the stable eras and the transitions are great opportunities for growing yourself**. Transitions are opportunities to raise the floor by building competency in new skills, and in stable periods you can raise the ceiling by developing mastery in the skills that the new era values. As the cycle repeats, your elevated floor will allow you to weather most transitions, and you'll thrive in most eras by leveraging your expanding masteries.

It's suggested that mature companies have more stable periods while startups have a greater propensity for change, but it's been my experience that what matters most is the particular team you join. I've seen extremely static startups, and very dynamic teams within larger organizations. I particularly want to challenge this old refrain:

"If you're offered a seat on a rocket ship, don't ask what seat! Just get on."
—Sheryl Sandberg

Even hypergrowth companies tend to have teams that are largely sheltered from change by either their management or because they're too far away from the company's primary constraints to get attention.

By tracking your eras and transitions, you can avoid lingering in any era beyond the point when you're developing new masteries. This will allow you to continue your personal growth even if you're working in what some would describe as a boring, mature company. The same advice applies if you're within a quickly growing company or startup: **don't treat growth as a foregone conclusion. Growth only comes from change, and that is something you can influence**.

6.2 Running a humane interview process

No matter how many times you've done it, changing companies is stressful because of the requisite job search and interviewing. Having conducted hundreds of interviews across a number of companies, I feel a bit more prepared to interview each time I do it, but being back on the interviewee's side of the table always leaves me humbled.

I'm confident that the state of interviewing is improving: many processes now involve a prepared presentation on a technical topic instead of an impromptu presentation (this more closely replicates a real work task), and many have replaced the whiteboard algorithm problem with a period of collaborative pair programming on a laptop with your editor of choice.

Looking back on my early interviewing experiences, when I was once asked to do calculus on the whiteboard, I'm amazed at how far things have improved.

That said, it's certainly not the case that interviewing has improved uniformly. There is still a lot of whiteboard programming out there, and a disproportionate number of the most desirable companies continue with the practice due to the combined power of inertia (it was the state of play when many engineers and managers—including myself—entered the profession) and coarse-grained analytics (if you're hitting your hiring goals—and with enough dedicated sourcers, any process will hit your hiring goals—then it can be hard to prioritize improving your process).

Reflecting on the interviews I've run over the past few years and those I got to experience recently, I believe that, while interviewing well is far from easy, it is fairly *simple*.

1. Be kind to the candidate.
2. Ensure that all interviewers agree on the role's requirements.
3. Understand the signal your interview is checking for (and how to search that signal out).
4. Come to your interview prepared to interview.
5. Deliberately express interest in candidates.
6. Create feedback loops for interviewers and the loop's designer.
7. Instrument and optimize as you would any conversion funnel.

You don't have to do all of these to be effective! Start from being nice and slowly work your way through to the analytics.

6.2.1 Be kind

A good interview experience starts with being kind to your candidate.

Being kind comes through in the interview process in a hundred different ways. When an interview runs overtime before you get to the candidate's questions, the kind thing to do is to allow the candidate a few minutes to ask questions, instead of running on to the next interview to catch up. Likewise, in that scenario the kind thing is to then negotiate new staggered start times, rather than kicking off a cascade of poor interviewer time management as each person tries to fractionally catch up to the original schedule.

My experience is that you can't conduct a kind, candidate-centric interview process if your interviewers are tightly time-constrained. Conversely, if an interviewer is unkind to a candidate (and these unkindnesses are typically of the "with a whisper not a bang" variety), I believe it is very often a structural problem with your interviewing process, and not something you can reliably dismiss as an issue with that specific interviewer.

Almost every unkind interviewer I've worked with has been either suffering from interview burnout after doing many interviews per week for many months or has been busy with other work to the extent that they have started to view interviews as a burden rather than a contribution. To fix that, give them an interview sabbatical for a month or two, and make sure that their overall workload is sustainable before moving them back into the interview rotation.

Identifying interview burnout is also one of the areas where having a strong relationship with open communication between engineering managers and recruiters is important. Having two sets of eyes looking for these signals helps.

6.2.2 What role is this, anyway?

The second critical step toward an effective interview loop is ensuring that everyone agrees on the role they are interviewing for, and what extent of which skills that role will require.

For some roles—especially roles that vary significantly between companies, like engineering managers, product managers, or architects—this is the primary failure mode for interviews, and preventing it requires reinforcing expectations during every candidate debrief in order to ensure interviewers are "calibrated."

I've found that agreeing on the expected skills for a given role can be far harder than anyone anticipates, and it can require spending significant time with your interviewers to agree on what the role requires. (This is often in the context of what extent and *kind* of programming experience is needed in engineering management, DevOps, and data science roles.)

6.2.3 Finding signal

After you've broken the role down into a certain set of skills and requirements, the next step is to break your interview loop into a series of interview slots that together cover all of those signals. Typically, each skill is covered by two different interviewers to create some redundancy in signal detection, in case one of the interviews doesn't go cleanly.

Just identifying the signals you want is only half of the battle, though; you also need to make sure that the interviewer and the interview format actually expose that signal. It really depends on the signal you're looking for, but a few of the interview formats that I've found very effective are:

1. Prepared presentations on a topic. Instead of asking the candidate to explain some architecture on the spur of the moment, give them a warning before the interview that you'll ask them to talk about a given topic for 30 minutes, which is a closer approximation of what they'd be doing on the job.

2. Debugging or extending an existing codebase on a laptop (ideally on *their* laptop). This is much more akin to the day-to-day work of development than writing an algorithm on the board. A great problem can involve algorithmic components without coming across as a pointless algorithmic question. (One company I spoke with had me implement a full-stack auto-suggest feature for a search inbox box, which required implementing a prefix tree, but the interviewer avoided framing it as yet another algo question.)

3. Giving demos of an existing product or feature (ideally the one they'd be working on.) This task helps them learn more about your product and get a sense of whether they have interest around what you're doing, and it helps you get a sense of how they deliver feedback and criticism.

4. Roleplays (operating off a script that describes the situation.) This option can be pretty effective if you can get the interviewers to buy

into it, allowing you to get the candidate to create more realistic behavior (architecting a system together, giving feedback on poor performance, running a client meeting, and so on).

More than saying that you should specifically try these four approaches (but you should!), the key point is to keep trying new and different approaches that improve your chance of finding signal from different candidates.

6.2.4 Be prepared

If you know the role you're interviewing for and know the signal which your interview slot is listening for, then the next step is showing up prepared to find that signal. Being unprepared is, in my opinion, the cardinal interview sin, because it shows a disinterest in the candidate's time, your team's time, *and* your own time. When I recall the fortunately rare situations when I've been interviewed by someone who was both rude and unprepared, I still remember the unprepared part first and the rude part second.

I've also come to believe that interview preparedness is much more company-dependent than it is individual-dependent. Companies that train interviewers (more on that below), prioritize interviewing, and maintain a survivable interview-per-week load tend to do very well, and other companies just don't.

Following from this, if you find that your interviewers are typically unprepared, it's probably a structural problem for you to iterate on improving and not a personal failing of your interviewers.

6.2.5 Deliberately express interest

Make sure your candidates know that you're excited about them.

I first encountered this idea reading Rands's "Wanted" article,[4] and he does an excellent job of covering it there. The remarkable thing is how few companies and teams do this intentionally: in my last interviewing process, three of the companies I spoke with expressed interest exceptionally well, and those three companies ended up being the ones I engaged with seriously.

Whenever you extend an offer to a candidate, have every interviewer send a note to them saying that they enjoyed the interview. (Compliment rules apply: more detailed explanations are much more meaningful.) At that point, as an interviewer, it can be easy to want to get back to your "real job," but resist the temptation to quit closing just before you close: it's a very powerful human experience to receive a dozen positive emails when you're pondering if you should accept a job offer.

6.2.6 Feedback loops

Interviewing is not a natural experience for anyone involved. With intentional practice you'll slowly get better, but it's also easy to pick up poor interviewing habits (asking brainteaser questions) or keep using older techniques (whiteboard coding). As mentioned earlier, even great interviewers can become poor when experiencing interview burnout or when they are overloaded with other work.

The fix for all these issues is to ensure that you build feedback loops into your process, both for the interviewers and for the designer of the interview process. Analytics (discussed in the next section) are powerful for identifying broad issues, but pair interviews, practice interviews, and weekly sync-ups between everyone strategically involved in recruiting (depending on your company's structure, this might be recruiters and engineering managers, or something else) work best to actively improve your process.

For *pair interviews*, have a new interviewer (even if they are experienced somewhere else!) start by observing a more experienced interviewer for a few sessions, and then gradually take on more of the interview until eventually the more senior candidate is doing the observing. Since your goal is to create a consistent experience for your candidates, this is as important for new hires who are experienced interviewing elsewhere as it is for a new college grad.

To get the full benefit of calibration and feedback, after the interview have each interviewer write up their candidate feedback independently before the two discuss the interview and candidate together. Generally, I'm against kibitzing about a candidate before the group debrief, in order to reduce bias in later interviews based on an earlier one, but I think this is a reasonable exception since you've experienced the same interview together. Also, in a certain sense, calibrating on interviewing at your company is about having a consistent bias in how you view candidates, independently of who on your team interviews them.

Beyond the interviewers getting feedback, it's also critical that the person who owns or designs the interview loop get feedback. The best place to get that is from the candidate and from the interview debrief session.

For direct feedback from candidates, I've started to ask every candidate during my "manager interview" sessions how the process has been and what we could do to improve. The feedback is typically surprisingly candid, although many candidates aren't really prepared to answer the question after five hours of interviews. (It's easy to get into the mode of surviving the interviews rather than thinking critically about the process that is being used to evaluate you.) The other, more common, mechanism is to have the recruiters do a casual debrief with each candidate at the end of the day.

Both of these mechanisms are tricky because candidates are often exhausted and the power dynamics of interviewing work against honest feedback. Maybe we should start proactively asking every candidate to fill out an anonymous Glassdoor review on their interview experience. That said, this is definitely a place where starting to collect some feedback is more important than trying to get it perfect in the first pass, so start collecting something and go from there.

6.2.7 Optimize the funnel

Once you have the basics down, the final step of building a process that remains healthy for the long haul is instrumenting the process at each phase (sourcing, phone screens, take-home tests, onsites, offers, and so on) and monitoring those metrics over time. If your ratio of sourced referrals to direct ones goes down, then you probably have a problem (specifically, probably a morale problem in your existing team). And if your acceptance rate goes down, then perhaps your offers are not high enough, but it also might be that your best interviewer has burned out on interviewing and is pushing people away.

Keep watching these numbers and listening to candidate post-process feedback, and you can go to sleep at night knowing that the process is still on the rails.

As a side note: I've put optimizing your funnel—and by this I include the entire process of building explicit analytics around your process—as the last priority in building a great interviewing process. From a typical optimization perspective, you should always measure first and optimize second, and here I'm giving the opposite advice.

Doing this first instead of last is definitely reasonable. In fact, I considered making this the first priority, and when I was setting up my last hiring process it was the first thing I did. In the end, I think you'll find that your process cannot thrive without handling the first six priorities, and that your analytics will direct you toward fixing those issues. Plus, the underlying data is very often poor, and it can be easy to get lost spending your cycles on instrumenting your process instead of improving it.

Altogether, the most important aspect to interviewing well is to budget enough time and to maintain a healthy skepticism about the efficiency of your current process. Keep iterating forward and your process will end up being a good one.

6.3 Cold sourcing: hire someone you don't know

The three biggest sources of candidates are *referrals* from your existing team, *inbound* applicants who apply on your jobs page, and *sourced* candidates whom you proactively bring into your funnel.

Small companies tend to rely on referrals, and large companies tend to rely on sourcing candidates, using dedicated recruiting sourcers for whom this is their full-time job (often the first rung on a recruiter career ladder). Medium-sized companies are somewhere on the continuum between those extremes. (Slack, in particular, has done some interesting work to encourage inbound applicants, and while they're now getting to the size at which most companies depend more heavily on sourced and direct applicants, I suspect that inbound applicants are not their largest source of candidates.)

Hiring and recruiting teams tend to prefer referrals because they often have higher pass and accept rates, and most early-stage companies, especially those without dedicated recruiting functions, end up being primarily composed of referrals. (An interesting caveat: lately I'm seeing more of a second category of referral candidate: those who are running their own extremely systematic interviewing process, with

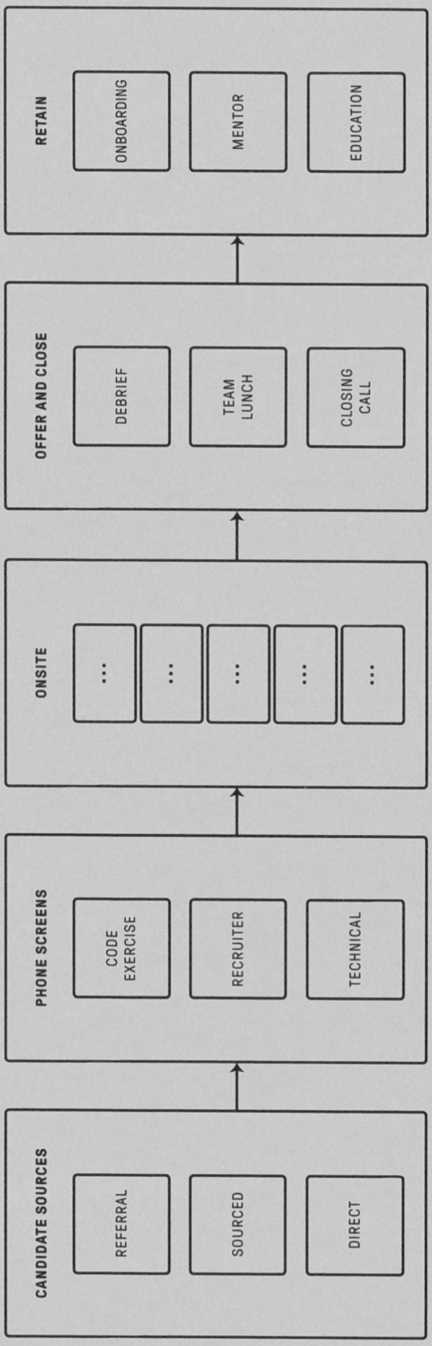

Figure 6.1
The phases of a hiring funnel.

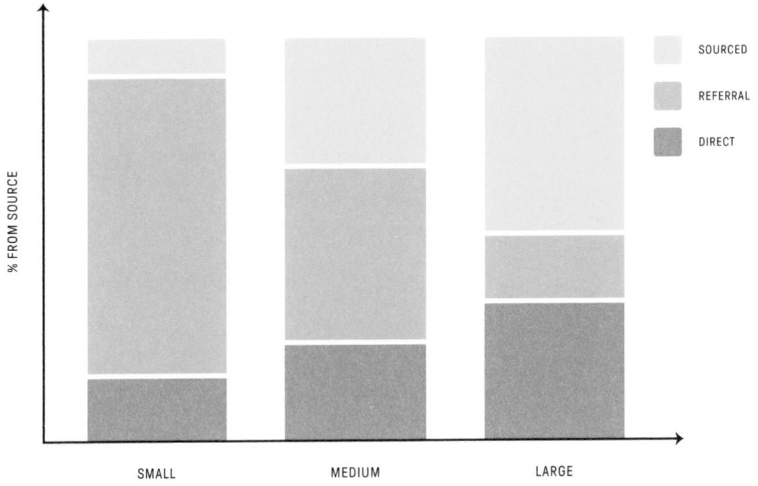

Figure 6.2
Small companies rely most on referrals, and larger companies on sourced candidates.

the aim of getting offers from three or more companies. Consequently, these folks tend to have much lower acceptance rates.)

Referrals come with two major drawbacks.

The first is that your personal network will always be quite small, especially when you consider the total candidate pool. This is especially true early in your career, but it's easy to work for a long time without building up a large personal network if you work in a smaller market or at a series of small companies. (One of the side benefits of working at a large company early in your career, beyond name recognition, is kickstarting your personal network.)

The other issue is that folks tend to have relatively uniform networks, composed of the individuals they went to school with or worked with. By hiring within those circles, it's easy to end up with a company whose employees think, believe, and sometimes even look similar.

6.3.1 Moving beyond your personal networks

Many hiring managers freeze up when their referral network starts to dry up, or as they look to bring a wider set of backgrounds onto their

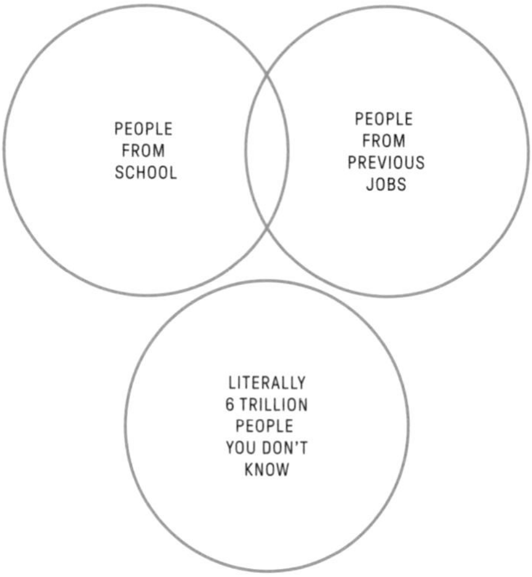

Figure 6.3
Small personal networks relative to total pool.

teams. However the good news is that there is a simple answer: cold sourcing. Cold sourcing, a technique that's also common in some kinds of sales, is reaching out directly to people you don't know.

If you're introverted, this will probably be an extremely unsettling experience at first, with questions like "What if they're annoyed by my email?" and "What if I'm wasting their time?" ringing in your head. These are important questions, and we have an obligation to be thoughtful about how we inject ourselves into others' lives. I was personally paralyzed by this concern initially, but ultimately I think it was unfounded: a concise, thoughtful invitation to discuss a job opportunity is an opportunity, not an infringement, especially for those who are on career networking sites like LinkedIn. Most folks will ignore you (which is great), others will politely demur (also great), a few will actually respond (even greater), and a surprising number will ignore you for six months and then pop up mentioning that they're starting a new job search. I've never had someone respond unkindly.

The other great thing about cold sourcing is that it's pretty straightforward. I'll share the approach I've used, with the caveat that I believe there are a tremendous number of different approaches that are probably more effective. Take this as a good starting point, track your results, and then experiment!

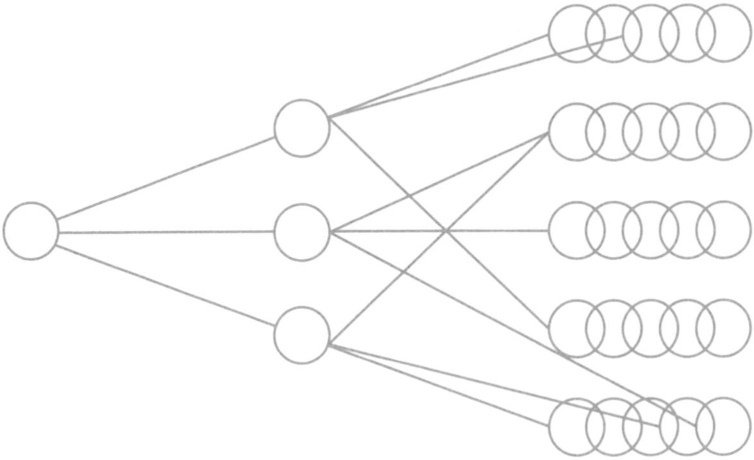

Figure 6.4
Three-degree network.

6.3.2 Your first cold sourcing recipe

My standard approach to cold sourcing is:

1. **Join LinkedIn**. I suspect variations of this technique would work on other networks (e.g., GitHub), but the challenge is that folks on other networks are generally not looking to engage about employment opportunities, and intent increases response rate significantly! A good parallel here is search versus display advertising, where search advertising has click-through rates that are an order of magnitude higher, as candidates are actually searching for what is being advertised.

2. Build out your network by **following individuals you actually do know**. Add everyone that you went to school with, have worked with, have interacted with on Twitter, etc. It's important to seed your network with some people you know because it'll increase the reach of your second-degree network, and it'll also reduce the rate at which folks mark you as someone they don't know (which is an input to being penalized as a spammer).

3. **Be patient**. If your initial network is small, it's very likely that you'll get throttled pretty frequently. Once you're throttled (you'll get a message along the lines of "You've exceeded your search capacity for this month"), you'll have to wait for a few days, potentially until the next month, to unthrottle. Alternatively, you could sign up for their premium products, which would accelerate this quite a bit. It may take weeks or months of occasional effort (schedule an hour each week) to get your network large enough that you're able to perform more than a few searches without rate limiting. Anecdotally, the number of connections at which things seem to get easier is around 600 or so.

4. Use the search function to identify **second-degree connections to connect to**. Start out by searching in your second-degree network by job title—*software engineer* or *engineering manager*—and as your network expands, consider switching from title to company. (Consider the various lists of great companies[5] in order to find companies to search by.) Build a broad church of connections! Even people whom you don't reach out to now are folks who might reach out to you later, or whom you might reach out to in a few months as your hiring priorities change. If you're not sure, just go ahead and do it.

5. When someone accepts your connection request, grab their email address from their profile and **send them a short, polite note** inviting them to coffee or a phone call, and sharing with them a link to your job description. Experiment with varying degrees of customization. (If you're having trouble finding their email, make sure you click *Show more* in their *Contact and Personal Info* section, and that they're a first-degree connection. A few folks don't share their email at all, and in that case I'd recommend moving on. Alternatively, you could send them a LinkedIn message directly.) I've personally found that customization matters less than I assumed, because people mostly choose to respond based on their circumstances, not on the quality of your note. (As a caveat, it's possible to write bad notes that discourage folks from responding. Iterating on your reach-out notes and getting a few other individuals with different perspectives to review your note is a quick and high-leverage thing to do.)

   Your reach-out notes can be very straightforward:

```
Hi $THEIR_NAME,

I'm an engineering manager at $COMPANY, and think
you would be a great fit for $ROLE (link to your
job description).

Would you be willing to grab a coffee or do a
phone call to discuss sometime in the next week?

Best,

$YOUR_NAME
```

Really, I think it can be that simple! If that seems *too* simple, then run an A/B test with something more personalized or sophisticated.

It's worth noting that you will end up connecting with some folks whom you simply don't have a great fit for right now. That's okay. I recommend reviewing their profile pretty rigorously after they accept your connection, and making an honest assessment on fit. If you don't have something, that's okay, and it's a better use of the candidate's time to not reach out to them. (However, I also think we tend to over-filter on qualities that don't matter too much! Being respectful of the candidate's time is, in my opinion, the most important thing to optimize for.)

6. Schedule and enjoy your coffees and chats, and remember that even people you aren't able to work with now are still folks you're likely to work with next year or next job. Especially in Silicon Valley, it's a very small network, and you should interact with each person as if they'll be providing feedback on whether or not to hire you at your next job. (They very well might!) You have two goals for each of these calls or coffees: figure out if there is a good mutual fit between the candidate and the role, and, if there is a good fit, try to get them to move forward with your process. The three things I find most useful to individuals deciding deciding to move forward with our process are describing why I personally am excited about the company and role we're discussing, explaining how our process works (from our initial chat all the way to them receiving an offer) and then leaving ample time for them to ask questions.

7. Keep spending an hour each week adding more connections and following up with folks who have connected. It's a bit of a grind at times, but it's definitely a practice that rewards consistency. I've found that this is a good activity to do together! Have a weekly meeting of coworkers who come together and source, chatting about how you're evolving your search. This also helps folks overcome their initial discomfort with cold reach-outs. (It's worth pointing out that this process is *much* easier to do with an applicant tracking system (ATS) like Lever[6] or Greenhouse,[7] which allow you to have a single place to track whether a candidate has already been contacted by someone else at your company. Having a bunch of folks from the same company reach out around the same time can paint a picture of chaos.)

If you've read through this and are quite confident that this approach won't work, I'm with you: before I tried it, I was similarly certain that it wouldn't work for me, and that it was just a big waste of time. But I've slowly been converted. It's also important to recognize that it's very likely this *exact* approach won't perform well over time. So try something simple, push through the concerns that block you from starting, and then experiment with different approaches.

6.3.3 Is this high-leverage work?

Similarly, a frequent follow-up question is whether sourcing is a high-leverage task for engineering managers. I think it is: candidates are more excited to chat with someone who'd be managing them than they are to chat with a recruiter whom they'll mostly work with during the interview process. Likewise, I think it's a valuable signal showing that managers care about hiring enough to invest their personal energy and attention into it.

That said, I would be cautiously concerned if an engineering manager was spending more than an hour a week on sourcing (not including follow-up chats, as those will take up a bunch more time). There is also a lot of important work to be done closing and evaluating candidates well, in addition to the numerous non-hiring-related opportunities to be helpful.

As a closing thought, the single clearest indicator of strong recruiting organizations is a close, respectful partnership between the recruiting and engineering functions. Spending some time cold sourcing is a great way to build empathy for the challenges that recruiters face on a daily basis, and it's also an excellent opportunity to learn from the recruiters you partner with! We've been doing weekly cold sourcing meetings as a partnership between our engineering managers and engineering recruiters, and it's been a great forum for learning, empathy-building, and, of course, hiring.

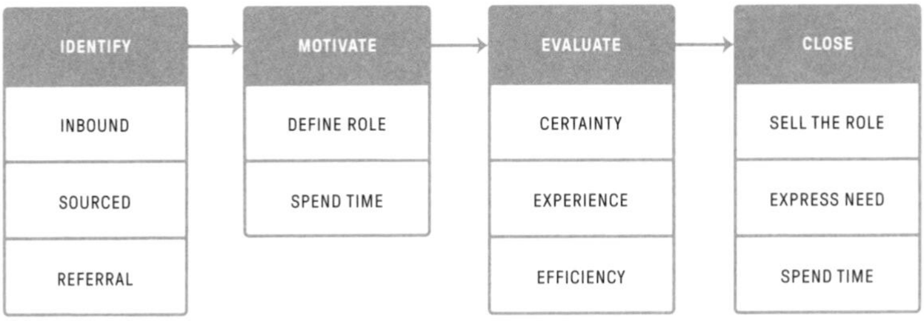

Figure 6.5
The phases of your recruiting pipeline.

6.4 Hiring funnel

Most companies believe that they are constrained by funding, product-market fit, or hiring. Books have been written about each of those, and this will be a foray into hiring. In particular, it'll be a look at how to use the fundamental hiring diagnostic tool: the hiring funnel.

6.4.1 Funnel fundamentals

The hiring funnel consists of four major steps: identifying candidates, motivating them to apply, evaluating them for your company, and closing them on joining. Depending on your particular circumstances, any or all of these can be very challenging.

⊙ Identify

Candidates tend to come from three large buckets: inbound, sourced, and referrals. Slower-growing and early companies tend to rely heavily on referrals, whereas fast-growing companies tend to exhaust their supply of referrals and to rely more heavily on sourcing and inbound.

- *Inbound* are candidates who apply to you directly. These come from your jobs website, often administered using an applicant tracking system like Greenhouse or Lever, or from job postings on LinkedIn and other job sites. Inbound tends to be high volume and low quality. The exception is for companies with powerful external brands, typically the culmination of strong product, reputation, and outreach.

- *Sourced* are candidates whom you proactively find and engage. The most common approaches are using LinkedIn, visiting colleges, and networking at conferences and meetups.

- *Referrals* are candidates whom someone at your company already knows, typically previous coworkers or friends who attended college together. This tends to be the primary source of hires for smaller companies. At most companies, referrals are the most efficient source of candidates, sporting the highest rate of interviews that lead to job offers.

⊙ Motivate

Once you've identified candidates you want to consider joining with your company, you need to motivate them to come interview! Some companies prefer to view this phase as a filter to weed out folks who are insufficiently passionate about their work, but I've not found that approach very effective. Rather, that approach seems to mostly filter for candidates willing to represent enthusiasm, as opposed to finding authentic passion. Instead, the formula that I've found most effective is pretty simple:

- *Spend time.* Have the people the candidate would work with spend time with them. Grab coffee with them, talk about the projects they're working on, and get them excited about learning from each other.

- *Clearly define the role.* Tell the candidate about what they'd be doing, being both very honest and a bit optimistic. That is to say, always give an accurate description of the work, but try to find the best frame for describing the work.

⊙ Evaluate

Once you're blessed with individuals who want to consider working with you, the next phase is to ensure that they'll be successful additions to your team. This phase is tricky because you're balancing quite a few objectives, some of which are in conflict:

- *Certainty.* You want to be as confident as possible that the candidate will be a success in your company. Letting employees go can be hard on morale, and it takes a great deal of time to do well.
- *Candidate experience.* You want candidates' motivation to join your company to increase as they're evaluated, not decrease. One of the worst possible outcomes of your hiring funnel is that you identify folks you want to join your company, but they're no longer interested in being part of it.
- *Efficiency.* You also want to minimize the amount of time invested by both your team and the candidate. How you think about this can lead to significant asymmetries, such as take-home assignments that require significant candidate time but little in-house time for evaluation (well, in principle anyway, since it seems like most folks find take-home assignments quite slow to review thoroughly).

⊙ Close

This is similar to the *motivate* phase, but now instead of asking them to commit a day, you're asking them to commit a couple years of their life. Many, many factors come in to play, from compensation packages to benefits, to making them feel needed. Because this is the final step, doing well here is an especially important factor in your funnel efficiency.

When you want to start operationalizing your hiring process, the first step is to craft a process for how candidates will flow through this funnel.

6.4.2 Instrument and optimize

Once you have your funnel defined, the second step is to instrument it! This is the most important argument for adopting a formal applicant tracking system, which will provide metadata around your process.

Instrumentation is so important because it allows you to understand where to focus your efforts. Different companies excel at different aspects, and even the same company will become better and worse at different stages over time. The only way to operate a consistently good hiring funnel is to ground your attention and efforts in your funnel metrics.

Once you do have the metrics, put your effort where there is the most room to improve. The first step of this is quite literal, looking at the conversion rates across each phase and investing. But what's slightly less obvious is what the reasonable upper bounds for each section should be. To answer those questions, benchmarking against your peer companies is really the only way to get useful information. If you don't benchmark, you can find yourself investing beyond diminishing returns in motivating candidates to interview, close, etc.

Whenever you have a hiring challenge, start from your instrumented hiring funnel and work the problem systematically from there!

6.4.3 Extending the funnel

The funnel we describe above is the most common variant, but I've found that a couple small tweaks make it much more powerful.

Instead of ending your funnel at closing candidates, add four more phases:

Onboard. How long does it take new hires to get up to speed? It's pretty tricky to measure this, but since you're trying to reason about the population rather than individuals, it's okay to be a bit messy. Pick a productivity metric, maybe commits per week, and see how long it takes for new hires to reach P40 productivity. That's a sufficiently good measure for understanding how long it takes folks to ramp up.

Impact. How impactful are the people that you're hiring? Again, that's a tricky one to measure, but you're looking to understand trends, not

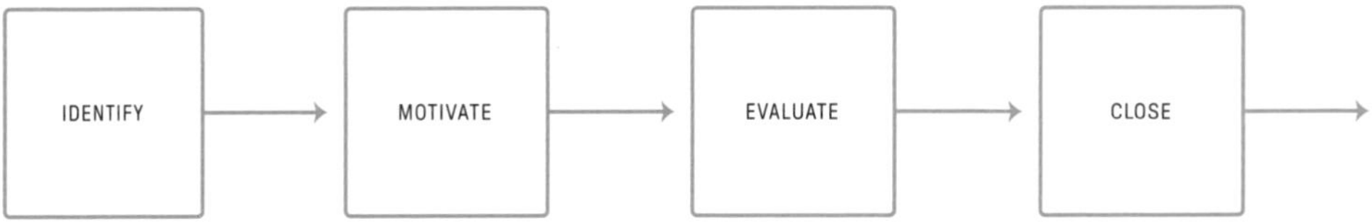

Figure 6.6
An extended hiring funnel, including onboarding, impact, and retention.

individuals, so don't worry about identifying perfect metrics. A reasonably good proxy here is looking at the performance rating distribution for new hires based on time since hiring.

Promote. How long does it take individuals to get promoted after they're hired? This is useful for understanding whether employees have access to opportunity[8] within your organization.

Retain. Are the people you hire staying? Fortunately, this one is pretty easy to track by looking at the people who leave. It is, however, quite a lagging indicator, since it typically takes years for folks to leave. Still, I believe that it's an essential metric to track.

Not many companies extend their hiring funnel this way, but I think it's quite useful, reframing hiring from a transactional process into your organization's lifeblood. It's also true that many companies are very uncomfortable sharing this kind of information widely. That's fine. These are indeed quite sensitive numbers, but you should make sure someone is paying attention to them.

6.5 Performance management systems

The most sacred responsibilities of management are selecting your company's role model, identifying who to promote, and deciding who needs to leave. At small companies these decisions tend to be fairly ad hoc, but as companies grow these decisions solidify into a formal performance management system. Many managers try to engage with these systems as little as possible, which is a shame. If you want to shape your company's culture, inclusion,[9] or performance, this is your most valuable entry point.

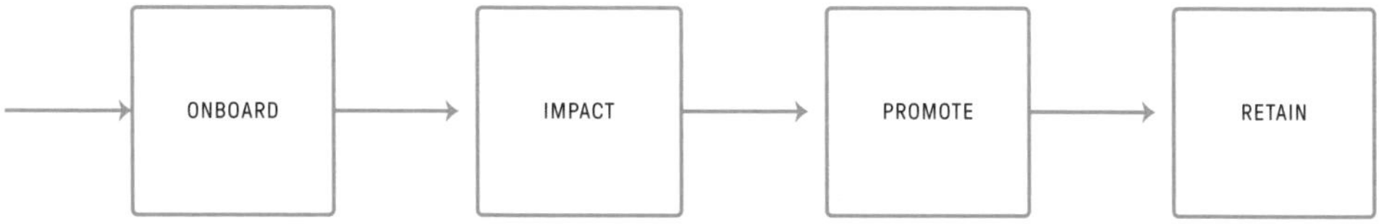

The number of approaches to performance management is uncountably vast, but most of them are composed of three elements:

Career ladders describe the expected evolution an individual will undergo in their job. For example, a software engineer ladder might describe expectations of a software engineer, a software engineer II, a senior software engineer, and a staff engineer.

Performance designations rate individuals' performance for a given period against the expectations of their ladder and level.

Performance cycles occur once, twice, or four times a year, with the goal of assigning consistent performance designations.

The purpose of these combined systems is to focus the company's efforts toward activities that help the company succeed. The output of these efforts is to provide explicit feedback to employees on how the company values their work.

6.5.1 Career ladders

At the foundation of an effective performance management system are **career ladders**, which describe the expected behaviors and responsibilities for a role. There is significant overhead to each career ladder that you write and maintain, and also a significant downside to attempting to group different roles onto a shared ladder.

What I've seen work best is to be tolerant of career ladder proliferation—really try to make a ladder for each unique role—but to only invest significant time in refining any given ladder as it becomes

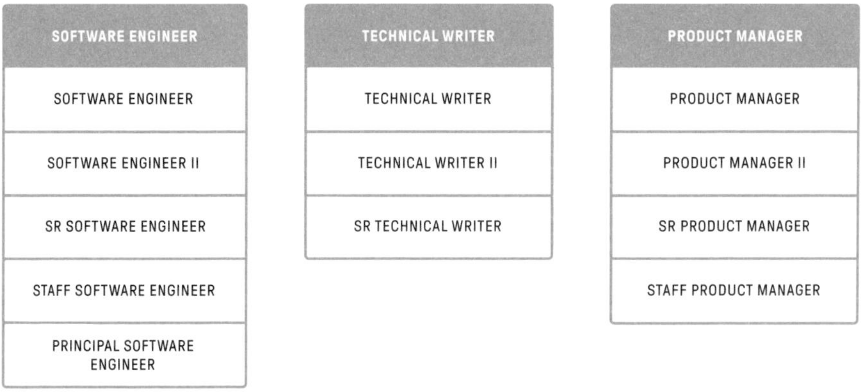

Figure 6.7
Diagram of a career ladder with multiple levels.

applicable for more employees. As a rule of thumb, any ladder with more than 10 individuals should probably be fully fleshed out, but smaller functions can probably survive with something rough. This works particularly well if you extend an open invitation to folks in that role to improve their ladder! (As an aside, I strongly recommend writing a lightweight ladder before you hire the first person into a given role. The alternatives tend to work out poorly.)

One effective method for reducing the fixed cost of maintaining ladders is to establish a *template* and *shared themes* across every ladder. Not only does this reduce fixed maintenance costs, it also focuses the company on a set of common values.

Each ladder is composed of **levels**, which describe how the role evolves in responsibility and complexity as the practitioner becomes more senior. The number of levels appropriate will vary across ladders, function size, and function age. Most companies seem to start with three levels and slowly add levels over time, perhaps adding one every two years. At each level, you'll want to specify the expectations across each of your values. Crisp level boundaries reduce ambiguity when considering whether to *promote* an individual across levels.

Crisp boundaries are also important as they provide those on a ladder a useful mental model of where they are in their journey, who their peers are, and whom they should view as role models. The level definitions are quite effective at defining the behaviors you'll want in your role models, which are the behaviors you'll see everywhere a year or two later.

A good ladder allows individuals to accurately self-access; these ladders are self-contained and short. A bad ladder is ambiguous and requires deep knowledge of precedent to apply correctly. If there is one component of performance management that you're going to invest into doing well, make it the ladders: everything else builds on this foundation.

6.5.2 Performance designations

Once you have the career ladders written, the next step is to start applying them. The most frequent application will be using them as a guide for self-reflection and during career coaching one-on-ones, but you'll also want to create formal feedback in the form of a **performance designation**.

A performance designation is an explicit statement of how an individual is performing against the expectations of their career ladder at their current level over a particular period of time. Because these designations are explicit, they are a backstop against miscommunication between the company and an employee. However it is a cause of concern and debugging if the explicit designation doesn't match the implicit signals someone has received.

Most companies start out using a *single scale* to represent performance designations, often whole numbers from one to five. Over time, these often move toward the *nine-blocker* format, a three-by-three grid with one axis representing performance and the other representing trajectory. Having used a number of systems, I prefer to use the simplest representation possible. The extra knobs in more complicated systems support more granularity, but my sense is that they simply create the impression of rigor while remaining equally challenging to implement in a consistent, fair way.

More important than the scale used for rating is how the ratings are calculated. The typical setup is:

1. **Self-review** is written by the individual receiving the designation. The best formats try to explicitly compare and contrast against their

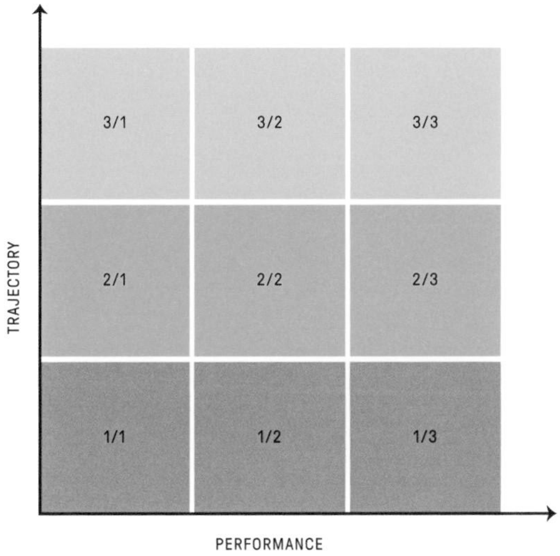

Figure 6.8
Diagram of a three-by-three grid, with one axis representing performance and another trajectory.

appropriate ladder and level. I've also seen good success in the "brag document"[10] format.

2. **Peer reviews** are written by an individual's peers, and are useful for recognizing mentorship and leadership contributions that might otherwise get missed. Structured properly, they are also useful for identifying problems that you're missing out on, but peers are generally not comfortable providing negative feedback.

3. **Upward reviews** are used to ensure that managers' performance includes the perspective of the individuals they manage directly. Format is similar to *peer review*.

4. **Manager reviews** are written by an individual's manager, typically a synthesis of the self-peer, and upward reviews.

From these four sets of reviews, you can establish a *provisional designation*, which you can then use as an input to a **calibration** system. Calibrations are rounds of reviewing performance designations and reviews, with the aim of ensuring that ratings are consistent and fair across teams, organizations, and the company overall.

A standard calibration system will happen at each level of the organizational tree. It's pretty challenging to strike the balance between avoiding calibration fatigue from many sessions and ensuring that the people doing the calibration are familiar with the work they are calibrating. Promotions are typically also considered during the calibration process.

Calibrations fall soundly in the unenviable category of things that are terrible but have no obvious replacement. Done poorly, they become bastions of bias and fierce politicization, but they're pretty hard to do well even when everyone is well-meaning! Some rules that I've found useful for calibrations:

1. **Adopt a shared quest for consistency**. Try to frame calibration sessions as a community of coworkers working together toward the correct designations. Steer them away from anchoring on the designations they enter with, and toward shared inquiry. Doing this well requires a great deal of psychological safety among calibrators, which needs to be cultivated long before they enter the room.[11]

2. **Read, don't present**. Many calibration systems depend heavily on whether managers are effective, concise presenters, which can become a larger factor in an individual's designation than their own work. Don't allow managers to pitch their candidates in the room, but instead have everyone read the manager review. This still depends on the manager's preparation, but it reduces the pressure to perform in the calibration session itself.

3. **Compare against the ladder, not against others**. Comparing folks against each other tends to introduce false equivalencies without adding much clarity. Focus on the ladder instead.

4. **Study the distribution, don't enforce it**. Historically, many companies fit designations to a fixed curve, often referred to as stack ranking.[12] Stack ranking is a terrible solution, but here's the problem it tries to address: it's easy for the meaning of a given designation to skew as a company grows. Instead of fitting to a distribution, I find it useful to review the distributions across different organizations and to discuss why they appear to be deviating. Are the organizations performing at meaningfully different levels, or have the meanings skewed?

Somewhat unexpectedly, performance designations are usually not meant to be the primary mechanism for handling poor performance.

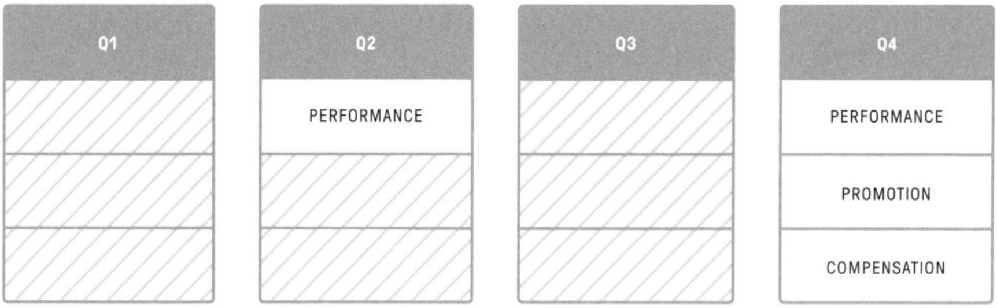

Figure 6.9
Diagram showing performance cycles happening in Q2 and Q4 but not Q1 or Q3.

Instead, feedback for weak performance should be delivered immediately. Waiting for performance designations to deal with performance issues is typically a sign of managerial avoidance. That said, it does serve as an effective backstop for ensuring that these kinds of issues are being addressed.

6.5.3 Performance cycles

Once you have career ladders and performance designations defined, you need a process to ensure that designations are being periodically calculated in a consistent and fair fashion. That process is your **performance cycle**.

Most companies do either annual or biannual performance cycles, although it's not unheard of to do them quarterly. The overhead of running a cycle tends to be fairly heavy, which leads companies to do them less frequently. Conversely, the feedback from the cycle tends to be very important, and it serves as a primary input into factors that individuals care about a great deal, like compensation, so there is also countervailing pressure to do them frequently.

The most important factor I've seen for effective performance cycles is forcing folks to practice. Providing well-structured timelines is very helpful, particularly if they're concise, but there tend to be so many competing demands that people do the most minimal skimming they can get away with.

Having teams do a practice round, at least for new managers or after the cycle has been modified, is the only effective way I've found to get around this. You can often direct this practice as an opportunity for folks to get feedback on their self-reviews, ensuring that they find it useful even if they're initially skeptical.

Finally, there is an interesting tension between improving the cycle as quickly as possible and allowing the cycle to stabilize so that people can get good at it. My sense is that you want to change the cycle at most once every second time. This lets everyone adapt fully, and it also gives you enough time to observe how well any changes work.

This is a small survey of some of the basics of designing performance management systems, and there is much, much more out there. It's valuable to start from the common structures that most companies adapt, but don't fall captive to them! Many of these systems are relatively recent inventions, and they take a particular, peculiar view of the ideal relationship between an employee and the company where they work.

If you're looking for more, Laszlo Bock's *Work Rules!*[13] is a good read.

6.6 Career levels, designation momentum, level splits, etc.

The fundamentals of performance management systems can be a bit bland! What's really interesting are the rough edges and unexpected emergent behaviors that come into play when you start to design and run these performance systems with lots of real people involved. These topics are particularly interesting to me because they are entirely unplanned yet they crop up consistently at pretty much every company as they scale.

Because these issues show up consistently, it's possible to be prepared for them, rather than being caught by surprise. Since surprise is the cardinal sin of performance management, they often create a bunch of trouble for managers earlier in their career, and hopefully these notes will help you navigate these confusing waters!

Designation momentum. This is the term for the natural tendency of a performance process to consistently produce the same evaluations for the same people despite changes in performance. If you are re-

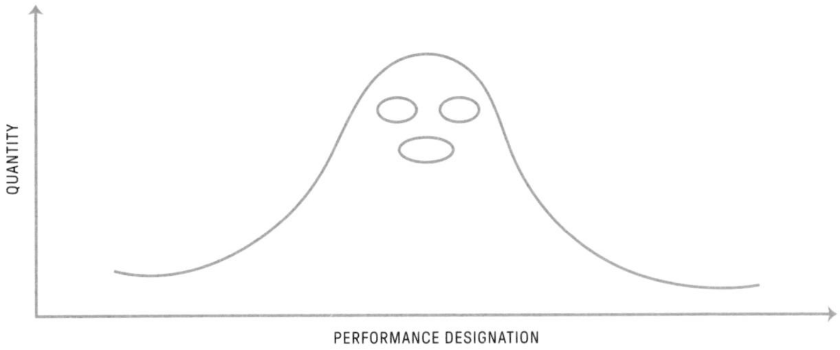

Figure 6.10
The relationship between Performance designation and quantity of designations.

ceiving good designations, this is an exciting phenomenon, because it means you are likely to continue receiving them. But I find that this is unexpectedly somewhat demotivating for high performers, who want consistent, direct feedback on their work so that they can keep improving. Those receiving poor designations, unsurprisingly, find this phenomenon quite frustrating, in particular because it makes it challenging for them to determine if it's a lagging indicator of a previous issue or if they're continuing to do poorly.

Many employees rely entirely on their manager to come up with a step-by-step path to high performance. That only works when designation momentum is taking you in a direction you're happy with. If it's not, you need to be the active participant in your success.

Propose a set of clear goals to your manager, and iterate together toward an explicit agreement on the expectations to hit the designation you're aiming for. The goals need to be ambitious enough that your manager can successfully pitch the difficulty to their peers during calibration. If your manager is pushing back on your goals' ambition, that is probably their way of saying that they think their peers will challenge their difficulty. That doesn't mean your plan isn't difficult enough—it may well be very appropriate—but it does mean you'll have to work to help them explain *why* the goals are appropriate.

Designation momentum occurs for individuals, but it also happens for teams and organizations. For teams in this position, setting clear goals is a good start, but it's also necessary to align with your peers and leadership about why your work is important. It's your work as a leader to explain why your work is important in terms that the organization understands and appreciates. This is a good example of where the failure to do so will have long-running costs.

Tit for tat. Calibration systems without strong process and fair referees can degrade into tit-for-tat favor trading. It's very rare to see active collusion during calibration; rather, the most common case occurs when folks are silent instead of raising concerns. This silence may seem benign, but it isn't: it pushes all responsibility for consistent outcomes onto the individuals refereeing the calibration process.

Encouraging engagement requires the calibration referee to model the behavior, but more importantly, it depends on building psychological safety and trust across the folks who are calibrating together.

Level expansion. As a company ages, it will inevitably add levels to support career progression. This happens even if a company remains the same size, and it's primarily driven by company age, not size. This is frequently driven by a small cohort of the highest-level executives.

If a company is experiencing particularly frequent *level expansion*, it is usually a sign that progression, compensation, or recognition has been overly tied to your level system, and you should identify mechanisms to reduce pressure on leveling. Training and education are useful here, as is getting more structured in assigning important projects.

The other scenario that typically leads to level expansion is one in which very senior executives from other, typically older, companies are hired. These people benefited from level expansion as that company aged, and it's hard for them to walk away from that heady cocktail of status, compensation, and recognition.

Level drift. Because *level expansion* is typically driven more by the need for career progression than by the introduction of objectively distinct accomplishment, levels added at the top create downward pressure on existing levels. Expectations at a given level decrease over time.

This inflation feels uncomfortable because we often rely on scarcity to determine value, but it's very uncommon for companies to adjust compensation in response to level drift. Thus, in practice, it is a rising tide that raises all boats. From a company perspective, it's important to manage level drift explicitly, so that it's possible to apply the shifts consistently.

Opening of the gates. The combination of *level expansion* and *level drift* leads to periodic bursts in which a cohort of individuals cross level boundaries together. This happens most frequently at the second-highest level, one or two cycles after a level expansion.

As a manager, you need to coordinate with your peers to ensure that you are opening the gate together in a consistent fashion. It's easy to miss these moments, but if you do you may inadvertently eject individuals from their natural cohort of peers. You can usually fix this in a subsequent cycle, but you'll have missed out on momentum. After each cycle, take an hour and try to guess when the gates might open next, and talk with your peers about it.

Career level. For every role, a given level will be established as the *career level*, and most individuals are not expected to progress beyond that level. Over time, this often leads to *career level clustering*, with the normal distribution centered on the career level, as opposed to the typical goal of the distribution centering at mid-level.

Time-at-level limits. Employees who haven't yet reached *career level* are expected to progress toward the career level at a consistent pace. This is typically used as a backstop for situations in which performance management seems appropriate but is not occurring. My experience is that most companies have time-at-level limits, but that there are many other ways to accomplish the same goals; such limits are useful as part of an overall system, but they aren't necessary in many configurations. The only bit I've found predictably important here is being consistent in how they are applied.

Level split. Over time, it is common for the *career level* to experience *level drift*, leading to increasingly distinct clusters of workers who reached career level at its highest expectations and those who have reached it recently. Given the greatly elevated expectations beyond the career level, upward mobility remains evasive. Many companies decide to perform a *level split*: separating the career level into two halves.

This allows the distinct cohorts to inhabit distinct levels, and it extends the runway of career progression for most employees. Less obviously, the split tends to solidify the moat guarding access to post-career levels. The extended moat doesn't catch those right on the border. It's easy to handle these folks properly, but the moat ab-

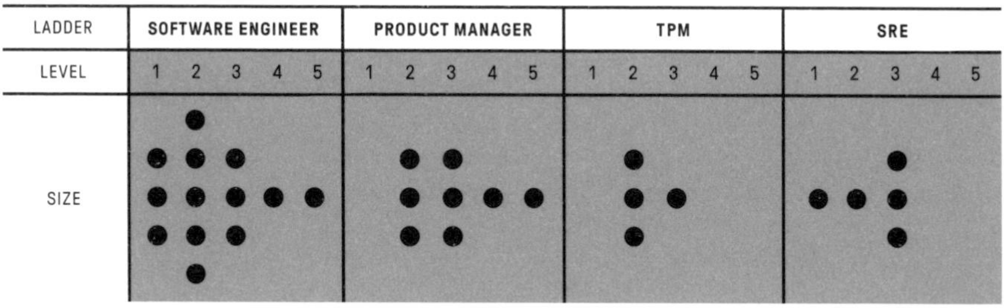

Figure 6.11
Distributions of levels across different functions.

solutely does slow the progression for the cohort who were about a year away from changing levels.

Crisis designations. These are alternatively known as *retention-driven designations*. Sometimes companies find themselves in a difficult situation, in which they have key individuals or even key teams that they consider to be at-risk, and one of the tools for addressing the situation is to recognize these individuals' importance through elevated performance designations. These are intended as temporary, but they tend to reset expectations permanently in ways that sacrifice long-term usefulness of the performance system in order to manage through short-term difficulty. Sometimes stuff gets really hard, and if that's the case, then use the tools at your disposal, but generally try to avoid doing this if possible.

There are, surely, hundreds more interesting topics when it comes to how performance systems work in practice as opposed to in design. Although these systems seem quite simple, I keep learning something new each time I go through a performance cycle, and I suspect that is a widely shared experience.

6.7 Creating specialized roles, like SRE or TPMs

People are sometimes surprised to learn that I started out working as a front-end engineer. I'd like to imagine it's because I'm so terribly knowledgeable about infrastructure, but I suspect that it's mostly grounded in my unconscionably poor design aesthetic. Something that has stuck with me from my front-end experience was feeling treated as a second-tier engineer: coworkers were unwilling to do any front-end work, but were careful to categorize it as trivial.[14] The following decade has seen radical improvements in browser compatibility and JavaScript tooling, and today's front-end engineers occupy an esteemed position in the hive mind's subtle hierarchy of roles.

While nodes have swapped positions, the hierarchy of roles remains alive and well, which is at its clearest when someone proposes creating a job description[15] or career ladder[16] for a new role. Most recently, the question of whether to create a dedicated career ladder for site reliability engineers[17] has been on my mind.

This particular question is dear to me, as I had the chance to design the initial iteration of Uber's SRE role, and while I think that the design was reasonably good, there are also so many ways it could have gone more smoothly. Faced with the decision of whether to do it a second time, my first instinct was to freeze and think of the ways it didn't work.

Grappling with the problem for some time, I remained conflicted, and decided to get more systematic around making this decision. I've written up the results of my musings here. Altogether, there are four interesting questions to dig into:

1. What are the pitfalls that these roles fall into?
2. If we do decide to create one, how do we set them up for success?
3. What are the benefits of specialized roles?
4. Putting it all together, when *should* you make a new role?

At the end, creating a new role will still be a difficult decision, but we'll be armed with a framework to help make it.

6.7.1 Challenges

The major challenges I've encountered rolling out new roles are:

Class systems. Folks often look at new roles as less important, framing them as service roles to absorb work they're not interested in. Sometimes roles are even explicitly designed this way, intended to reduce work for another role as opposed to having an empowering mission of their own.

Brittle organization. As you move away from generalized roles and toward specialists, an unexpected consequence is that your organization has far more single points of failure. Where everyone on a team was once able to perform all tasks fairly effectively, now if your project manager leaves, you'll find that no one is able to fill the role very capably. This brittleness is particularly acute in organizations with frequent structural changes.[18]

Pattern matching. Designing a new role for your organization tends to involve dozens of important decisions in order to align it with your needs. Unfortunately, folks generally don't take much time to appreciate these distinctions, and instead pattern match on how they've seen the role done elsewhere. This is a powerful force. Some meaningfully large percentage of people will both avoid taking any steps to learn how the role is intended to function—reading documentation, asking about the approach—and continue to express surprise that it doesn't work exactly the way they saw at a previous company.

Task offloading. When a new role is created, the role's designers have a very clear vision of how they want the new function to work. Many other individuals are not particularly concerned with how the creators want the function to work, and will view it as an opportunity to offload tasks that they find challenging, difficult, or uninteresting. This can lead to new roles being immediately underwater, which often feels like success to leaders attempting to grow the size of their organization. However, that can can easily translate into an unlovable work experience for those performing the role.

Roles too "trivial" to value. Many roles start by taking on work that is viewed as uninteresting by the role shedding that responsibility, and consequently individuals in that role tend to view that work as trivial and unimportant. This often translates into the new roles struggling to have their impact be recognized.

Roles too "trivial" to promote. Similar to the above, you'll often find that the work done by new roles is valued very highly in terms of impact, but not viewed as sufficiently "strategic" to merit promotion, particularly beyond career level.[19] This can lead to individuals being obligated to change roles if they want to attain higher tiers of achievement.

Head count obstacles. Companies eventually develop arcane rituals by which a series of emails, meetings, and incantations is translated into an annual head count plan. These systems are, quite reasonably, designed around supporting the needs of large, existing roles. Consequently, they tend to make it pretty challenging to get head count for a new role, particularly if it's in tension with existing functions that need to expand.

Recruiting rare humans. For entirely great reasons, people want the first hires they make into a new role to be strong role models for the entire function. This often leads to a proliferation of requirements until it's impossible for any candidate to pass the bar.

Inability to evaluate. This is almost the opposite of the above challenge: sometimes the existing organization has so little experience with the new function they wish to create that they simply don't have a usable means by which to evaluate candidates. This can lead to evaluations focused on qualities that are largely independent of what the candidates would be doing if they accepted the job.

6.7.2 Facilitating success

If we do want to create a new role, it's important to take stock of what we'll need to do to make employees in the role successful. The ingredients that I've found necessary for a new role to bootstrap successfully are:

Executive sponsor. Not necessarily an executive, but you'll need to find an authorized, senior leader who is committed to the success of the new function. Especially for the first performance[20] and head count cycles, there will be a number of rough edges that will require significant organizational support to navigate. The need to find a sponsor who'll be able to provide the necessary support is the most obvious constraint on creating new roles. If you can't find a sponsor, it's usually important feedback that leadership doesn't believe the new role will have a good return on invested energy.

Recruiting partner. A new role will require significant support from your recruiting organization in order for it to succeed. Every role being recruited against has a high fixed cost, and adding new roles can make it difficult for recruiters to hit the targets that their performance is measured against. Make sure that your recruiting team is able to support a new role! If they aren't, the first step may be working with your executive sponsor to direct more head count toward recruiting.

Self-sustaining mission. New roles are frequently described in terms of how they'll impact other functions, rather than in terms of what they'll accomplish. For example, you might describe technical program managers (TPMs) as offloading project management responsibilities from engineering managers. This approach frames the role as an auxiliary, support function, which makes it difficult to recognize the work's impact. You must be able to frame the role's work without referencing other existing roles in order for it to succeed long-term.

Career ladder. In pretty much all cases, new roles should have a career ladder from the beginning. The career ladder is the foundation of a successful performance management system, and it's not possible for a role to be valued or evaluated coherently without a thoughtful career ladder. Sometimes folks rush ahead to hire before writing the ladder, but the work required to design an effective interview loop is roughly equivalent to writing a career ladder, so I've found that skipping this step is an act of false economy.

Role model. Who will be the external and internal role models for this role? A good role model is a human embodiment of the intent codified in your career ladder. You want to have a person you can point to.

Dedicated calibrations. Most performance management systems rely on a calibration system to ensure that performance designations are being assigned in a consistent fashion across teams and roles. Sometimes managers try to perform calibration with mixed roles in a single session, which leads to smaller roles being treated as afterthoughts. Often the designations will be approved without much thought, or they'll be pushed toward the center. Neither of those scenarios creates a useful feedback loop for the individual in the new role. It's best to consider them separately in a dedicated calibration session for the one role. If that's not possible, the second-best option is to consider all smaller roles together, so that various forms of specialized contributions can be considered thoughtfully.

6.7.3 Advantages

If creating a new role was all costs and challenges, it would be easy to decide not to move forward, but there are pretty significant advantages to be had:

Efficiency. This is the flip side of *brittle organization*: people in specialized roles are able to spend more time doing a smaller set of tasks, which leads to great expertise in that area. For areas where existing roles are spending significant amounts of time, this specialized efficiency can translate into significantly more overall throughput without increasing head count. I think this is the most important advantage, and it's especially valuable for teams or companies in which financial resources are the limit on growth (which is most teams at most companies).

Efficiently solve constraints. This is an extension of the efficiency point, but subtly different: with specialists, you can add exactly the kind of capacity you are missing, which is very powerful for efficiently solving constraints. If your organization is low on project management bandwidth, you could add five new managers who can each take on a bit of it, or you might be able to add a single project manager who individually adds as much relevant bandwidth as the five managers combined.

Recognition. Simply creating a new role will absolutely not cause folks to suddenly start valuing work that they previously dismissed, but it can be a useful component in that shift. In particular, it will provide additional structural mechanisms to support recognition, such as distinct career ladders, calibration sessions, and even compensation structures.

Evaluating for strengths. It's often challenging to interview specialists effectively, because you'll typically evaluate them based on how they'd perform for your generalist position, while missing out on their peculiar areas of excellence. Creating a new role makes it possible to target the interview process on the areas that are most important.

Increased hiring pool. You're now able to consider a new pool of candidates in your hiring funnel,[21] and that potentially expands the total number of candidates you can consider.

Specialized compensation. In some cases, the market compensation for specialists is significantly higher than that for generalists, and in that case it's often quite a challenge to hire specialists without specialized compensation bands.

6.7.4 What to do?

Once you're familiar with the challenges and the costs of provisioning a new role, all that's left is to consider the advantages and make a quick judgment call. Ah, well, actually it's still a pretty hard decision to make!

Some questions that can guide this decision are:

- Is there a less extreme way to address the recognition gap? Maybe you could adjust the existing career ladder instead.

- Do you have a plan for changing how the company values the work? Creating a new role won't inherently change how the company values this work. You'll still need to do the hard work of expanding your company's values.

- If you have a plan to change company values, could you do a trial run of that plan before introducing the new role? This helps de-risk the experiment for the folks you are trying to help, and it's much easier to boot.

- Does the function already exist in secret? Sometimes you'll find that roles have already split, and you're less making a new function than recognizing an existing one.

- Will this increase short-term recognition of performance but ultimately hurt the career growth of employees who change roles? Creating a new role is absolutely the kind of thing that can initially feel like progress but ultimately sets folks back significantly, requiring them to transition to other roles later.

- Is the number of people impacted sufficiently high and is the recognition gap of value significantly large, to cover the sizable costs of creating and nurturing a new role?

- Who will pay the maintenance costs for the new role? If the answer is that you'll personally pay them, who will take up the torch if you leave?

As you think through those questions, hopefully the right approach for your situation gets a bit clearer. As a rule of thumb, I would always create a new role if it immediately covered 20 individuals; would reluctantly create a new role if it would cover 20 individuals within two years; and would be skeptical of creating a new role that couldn't meet one of those two conditions!

6.8 Designing an interview loop

Anyone who has flipped through *Cracking the Coding Interview* by Gayle Laakmann McDowell[22] knows that evaluating candidates for a new role is a coarse science. Most interviewers are skeptical of the accuracy of their interviews, and it's hardly the rare interview retrospective where interviewers aren't sure they got enough signal on a candidate to hire with confidence.

Although a dose of caution about interview accuracy is well-founded, I'm increasingly convinced that a bit of structure and creativity leads to interview loops that give a clear signal on whether a candidate will succeed in the role, and whether they can do so in a fair and consistent fashion.

The approach I've started using to design interview loops is:

Metrics first. Do not start designing a new interview loop until you've instrumented your hiring funnel.[23] It's pretty hard to evaluate your loop or improve it without this data, so don't undertake an improvement project without it.

Understand the current loop's performance. Take time to identify what you think does and doesn't work well in your current process. Three sources that are useful are:

- Funnel performance. Where do people drop off in your funnel? How does your funnel benchmark against your peers at similar companies?
- Employee trajectory. For all the candidates you have hired, understand how their work performance relates to their interview performance. What elements correlate heavily with success, and which appear to filter on things that end up not contributing to performance in the job?
- Candidate debriefs. Try to schedule calls with everyone who goes through your process, especially with those who drop out.

Learn from peers. Chat with folks who have interviewed at other companies for the kind of role you're hoping to evaluate. You're not

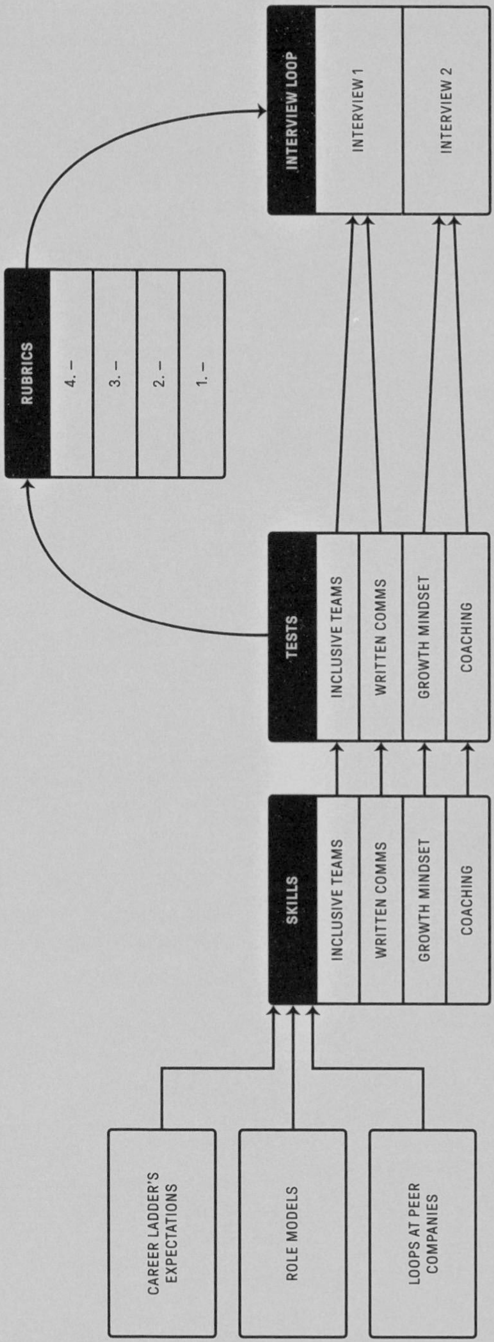

Figure 6.12
Steps of designing an interview loop.

looking to copy elements verbatim, but rather to get a survey of existing ideas.

Find role models. Write up a list of ideal candidates for the role and write down what makes them ideal. Be deliberate about ensuring that your list of role models includes women and underrepresented minorities, helping avoid matching on correlating traits in a less diverse set of role models.

Identify skills. From your role models and your career ladder,[24] identify a list of skills that are essential to the candidate's success in the role they'll be interviewing for. Take a bit of time to rank those skills from most to least important.

Test for each skill. For each skill, design a test to evaluate the candidates' strengths. Whenever possible, prefer tests that have a candidate show their strengths, avoiding formats in which they tell you about it. For example, we took an interview in which folks described their experience building a healthy team, and replaced it with an interview in which folks are given results from an organization health survey and asked to identify issues and propose how they'd address them.

- Avoid testing for polish. Many interviews accidentally test for the candidate's polish as opposed to testing for any particular skill. This is especially true for experiential interviews in which folks are asked to describe their work, and less common in interviews that ask them to demonstrate skills. That isn't to say that you shouldn't deliberately test for polish—it's quite useful—just that you shouldn't do it inadvertently.

For each test, a rubric. Once you've identified your tests, write a rubric to assess performance on each test. Good rubrics include explicit scores and criteria for reaching each score. If you find it difficult to identify a useful rubric for a test, then you should look for a different test that is easier to assess.

- Avoid Boolean rubrics. Some tests tend to return Boolean results: for example, whether someone has experience managing someone is a good example of a common Boolean filter. These are inefficient tests because you pretty quickly get a sense of whether someone has or hasn't ticked this box, and the rest of the interview doesn't lend more signal. Likewise, you can almost always answer Boolean questions from an applicant's resume or in a pre-interview screen.

Group tests into interviews. Once you've identified the tests, group them into sets that can be performed together in a single 45-minute interview, or whatever duration you prefer. The more cohesive the format and subjects of the tests in a given interview, the more natural it'll feel for the candidate.

Run the loop. At this point you have an entire loop ready, and it's time to start using it. Especially early on, you should be asking candidates what did or didn't work well—but, really, you should never stop doing this! Each debrief will uncover some opportunities to improve your rubric or tests.

Review the hiring funnel. After you've run the new loop 10 to 20 times, review the funnel metrics to see how it's working out. Are some of the interviews yielding too many passes? Are others too hard? Review the results in batch.

Schedule an annual refresh. As the initial rate of iteration slows down, schedule a review for a year out, and at that point you get to ask yourself if the loop has proven effective for your needs, or if you should restart this process!

At this point, you have a complete interview loop and the systems to guide the loop toward improvement. Beyond this approach, here are a few more general pieces of guidance:

Try to avoid design by committee. These almost always lead to incremental change. Prefer a working group of one or two people that is then tested against a bunch of candidates for feedback!

Don't hire for potential. Hiring for potential is a major vector for bias, and you should try to avoid it. If you do decide to include potential, then spend time developing an objective rubric for potential, and ensure that the signals it indexes on are consistently discoverable.

Use your career ladder. Writing a great interview loop is almost identical to writing a great career ladder.[25] If you've already written expectations for the role, reuse those as much as possible.

Iterate on the interview a little. When you first create an interview, you should spend time iterating on the interview format, but the rate of change should drop to near-zero after you've given it approximately 10 times.

Iterate on the rubric a lot. As you attempt to apply the interview rubrics, you'll continue to find edge cases and ambiguities. Continuing to incorporate those into the interview rubrics is an essential way to reduce bias creeping in.

A/B testing loops. There is a platonic ideal of testing new interviews using the standard mechanisms we use to test other changes, such as A/B testing. At a certain scale, I think that is almost certainly the best approach, but so far I've been part of a company with enough volume to usefully conduct such tests. In particular, these tests are quite expensive, because you need to control for interviewers and have those interviewers trained on both sets of interviews, which ends up taking a long time to reach confidence that a new loop is better.

Hiring committees. As an alternative to A/B testing, I have found a centralized hiring committee that reviews every candidate's interview experience to be quite valuable for identifying trends across new loops. More generally, this approach helps guide the overall process toward consistency and fairness.

Wrapping all my learnings on designing interview loops together into something pithy: avoid reusing stuff that you know doesn't work, and instead approach the matter from first principles with creativity. Then keep iterating based on how it works for candidates!

7

CHAPTER 7

Appendix

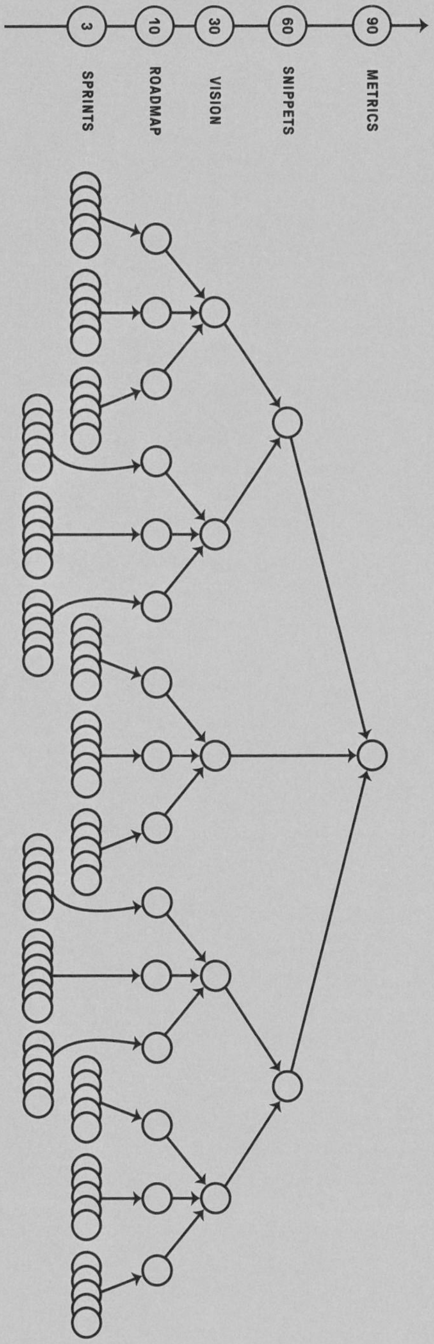

Figure 7.1
Organization processes added to support operating it as it scales.

7.1 Tools for operating a growing organization

One tension in management is staying far enough out of the details to let folks innovate, yet staying near enough to keep the work well-aligned with your company's value structures. Seeing this challenge from a variety of perspectives with different teams, I've collected a playbook of tools to facilitate the balancing act, and a loose framework for rolling those tools out.

A few caveats about process rollouts:

1. Teams and organizations have a very limited appetite for new process; try to roll out one change at a time, and don't roll out the next change until the previous change has enthusiastic compliance.
2. Process needs to be adapted to its environment, and success comes from blending it with your particular context.

7.1.1 Line management

Around the time your team reaches three engineers, you'll want to be running a **sprint process**. There are many successful ways to run sprints. Try a few and see what resonates for you.

The criteria I use to evaluate if a team's sprint works well:

1. *Team* knows what they should be working on.
2. *Team* knows why their work is valuable.
3. *Team* can determine if their work is complete.
4. *Team* knows how to figure out what to work on next.
5. *Stakeholders* can learn what the team is working on.
6. *Stakeholders* can learn what the team plans to work on next.
7. *Stakeholders* know how to influence the team's plans.

One of my coworkers, Davin Bogan,[1] likes to say that "shipping is a habit," and a well-run sprint both helps teams establish that habit and

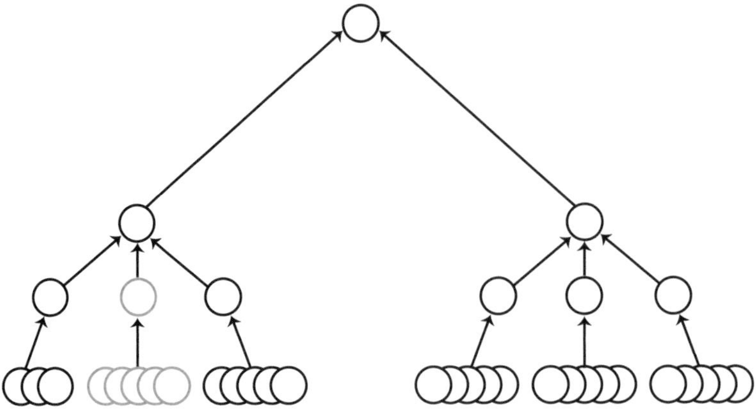

Figure 7.2
Org chart for a line manager.

serves as a mechanism that creates visibility within a team that hasn't quite gotten there yet. As a team's direct manager, you can use this to ground concerns around individuals who might not be ramping up successfully, and, as you move into middle management, sprints are useful for debugging within your organization.

Within your sprint process, your **backlog** is particularly important: it's the context-rich interface that you'll use to negotiate changes in direction and priority with your stakeholders. It's always more interesting to discuss which of two things we should do next, rather than whether something is worth doing.

As your team gets larger and the number of stakeholders you're working with grows, you'll also want to develop a **roadmap** describing your high-level plans over the next three to twelve months. Planning does not inherently create value, so aim to keep your roadmap as short as possible and allow teams to coordinate.

Initially, the distinction between your backlog and your roadmap may be quite small: your backlog a bit more detailed, your roadmap looking a bit further into the future. The value in having both is that this lets you specialize the backlog to be more useful for your team and design the roadmap to be more useful for your stakeholders, rather than relying on one tool to satisfy both sets of constraints.

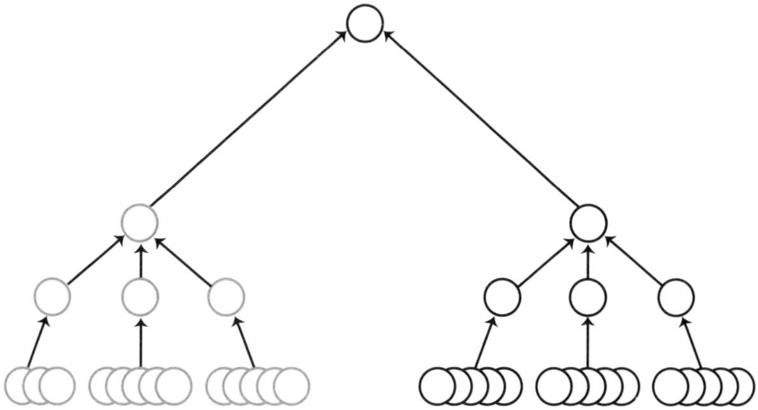

Figure 7.3
Org chart for a middle manager.

At this point, most teams will be tracking **operational metrics**, with a focus on tracking day-to-day user and system behavior. These metrics tend to be exclusively focused on helping the team operate, and in particular detect outages, regressions, and other interruptions.

7.1.2 Middle management

As you move into middle management, you'll become responsible for two to five line managers. As a result, you'll need to shift away from day-to-day execution to give your line managers room to make an impact (and in order to free yourself up to make a larger impact as well).

You'll be spending more time on your *roadmaps* as:

- Your move from receiving asks from stakeholders to deeply understanding what is motivating those asks.
- You invest in learning what other folks are working on in order to continuously validate that your teams' efforts are valuable.

As you spend less time with your teams, you'll want to start a **weekly staff meeting** with your managers. The best versions I've seen start with brief updates from each attendee, at most a couple of minutes per person, and then move into group discussions on shared topics.

Topics might include running effective sprints, planning, career development, or whatever else proves useful. Done well, these discussions are the key learning forum for you and the managers you work with.

As your teams and the organization around you grow, you'll start to see more and more cases of preventable misalignment: two teams working on similar projects because they're unaware of each other, another team struggling because they don't have a reliable email service when your team *actually does* have one. At that point, it's time for each team to write a **vision document**: a concise statement of the team's goals and the strategy for accomplishing those goals.

It'll be extremely frustrating for some teams to write their vision documents, because it'll force them to recognize and reconcile areas of distributed and unclear ownership. It's worth the pain! Once your vision comes together, it becomes your roadmap's North Star, and it will help you reconcile stakeholder asks with your longer-term product and technology strategies.

This is also when you'll want to **start skip-level one-on-ones** to ensure that there are direct, open channels for feedback about your managers and your teams. Typically, if you're learning something negative during a skip-level, you should have learned it somewhere else first, but rigorous processes have some redundancy. Nothing works consistently every time.

7.1.3 Managing an organization

As your organization starts to get even larger and you're mostly managing middle managers, the playbook shifts again. Your staff meeting has changed in one of two ways:

- The meeting has so many managers in it that they can't even provide important updates. Plus the discussions have become unwieldly, with a couple of folks dominating conversations.

- Alternatively, your meeting now includes your middle managers, who themselves are likely missing some of the context about what their teams are working on or struggling with.

The mechanism I've found most helpful in this case is to ensure that every team has a clear set of **directional metrics** in an easily discov-

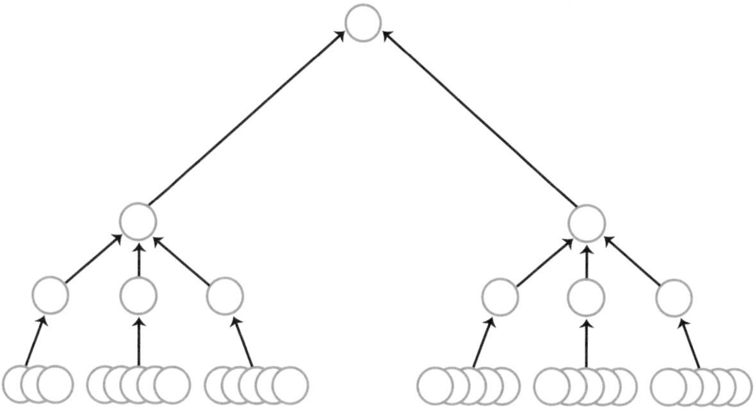

Figure 7.4
Org chart for a director.

erable dashboard. The metrics should cover both the longer-term goals of the team (user adoption, revenue, return users, etc.) and the operational baselines necessary to know if the team is functioning well (on-call load, incidents, availability, cost, and so on.) For each metric, the dashboard should make three things clear: the current value, the goal value, and the trend between them.

Now your staff meetings can start with a quick metrics review to give a sense of whether there is somewhere you need to drill in, and, rather than peanut buttering, you can focus your attention on projects that are exceeding or struggling.

The other mechanism I've found to be exceptionally useful at this point is **team snippets**. These come out every two to four weeks and give snapshots of each team's sprints: what they're doing, why they're doing it, and what they're planning to do next. These are *valuable* for you to retain a sense of what your teams are working on, but they are *invaluable* for decentralizing coordination and communication between teams in your organization, as you become increasingly ineffective in that role.

Along the way, remember that your old problems still exist, it's just that other folks are dealing with them instead. As you roll out new processes to solve your personal pain points, you should be handing off processes to your managers, and keeping those practices intact and running. This will leave you with a tapestry of reinforcing processes, which support you and each layer of management that you support.

7.2 Books I've found very useful

Folks occasionally ask me to recommend books to help them in their professional career. I can usually think of a couple recommendations in the moment, but I always feel as if I'm forgetting far more good books than I'm recommending. In the hope of providing a better answer going forward, I've written up some of the general purpose, leadership, and management books I've read.

Not all of these are classically great books—some are even a bit dull to read—but they've changed how I think in a meaningful way. They're roughly sorted in order from those I found most valuable to least:

1. ***Thinking in Systems: A Primer***
   by Donella H. Meadows

   For me, systems thinking has been the most effective universal tool for reasoning through complex problems, and this book is a readable, powerful introduction.

2. ***Don't Think of an Elephant! Know Your Values and Frame the Debate***
   by George Lakoff

   While written from a political perspective that some might find challenging, this book completely changed how I think about presenting ideas. You may be tempted to instead read Lakoff's more academic writing, but I'd recommend reading this first as it's much briefer and more readable.

3. ***Peopleware: Productive Projects and Teams***
   by Timothy Lister and Tom DeMarco

   The book that has given generations of developers permission to speak on the challenges of space planning and open offices. Particularly powerful in grounding the discussion in data.

4. ***Slack: Getting Past Burnout, Busywork, and the Myth of Total Efficiency***
   by Tom DeMarco

   Documents a compelling case for middle managers as the critical layer where organizational memory rests and learning occurs. A meditation on the gap between efficiency and effectiveness.

5. ***The Mythical Man-Month***
   by Frederick Brooks

   The first professional book I ever read, this one opened my eyes to the wealth of software engineering literature waiting out there.

6. ***Good Strategy/Bad Strategy The Difference and Why it Matters***
   by Richard Rumelt

   This book gave me permission to acknowledge that many strategies I've seen professionally are not very good. Rumelt also offers a structured approach to writing good strategies.

7. ***The Goal: A Process of Ongoing Improvement***
   by Eliyahu M. Goldratt

   Explores how constraint theory can be used to optimize process.

8. ***The Five Dysfunctions of a Team***
   by Patrick Lencioni

9. ***The Three Signs of a Miserable Job***
   by Patrick Lencioni

   Another Lencioni book, this one explaining a three-point model for what makes jobs rewarding.

10. ***Finite and Infinite Games***
    by James P. Carse

    Success in most life situations is about letting everyone continue to play, not about zero-sum outcomes. This seems pretty obvious, but for me it helped reset my sense of why I work.

11. ***INSPIRED: How to Create Tech Products Customers Love***
    by Marty Cagan

    A thoughtful approach to product management.

12. ***The Innovator's Dilemma: When New Technologies Cause Great Firms to Fail***
    by Clayton M. Christensen

    A look at how being hyper-rational in the short run has led many great companies to failure. These days, I think about this constantly when doing strategic planning.

13. ***The E-Myth Revisited: Why Most Small Businesses Don't Work and What to Do About It***
    by Michael E. Gerber

    The idea that leadership is usually working "on" the business, not "in" the business. Work in the business to learn how it works, but then document the system and hand it off.

14. ***Fierce Conversations: Achieving Success at Work and in Life, One Conversation at a Time***
    by Susan Scott

    How to say what you need to say. This is particularly powerful in giving structure to get past conflict aversion.

15. ***Becoming a Technical Leader: An Organic Problem-Solving Approach***
    by Gerald M. Weinberg

    Permission to be a leader that builds on your strengths, not whatever model that folks think you should fit into.

16. ***Designing with the Mind in Mind***
    by Jeff Johnson

    An introduction to usability and design, grounding both in how the brain works.

17. ***The Leadership Pipeline: How to Build the Leadership Powered Company***
    by Ram Charan, Steve Drotter, and Jim Noel

    This book opened my eyes to just how thoughtful many companies are in intentionally growing new leadership.

18. ***The Manager's Path: A Guide for Tech Leaders Navigating Growth and Change***
    by Camille Fournier

19. ***High Output Management***
    by Andy S. Grove

20. ***The First 90 Days: Proven Strategies for Getting Up to Speed Faster and Smarter, Updated and Expanded***
    by Michael D. Watkins

21. ***The Effective Executive: The Definitive Guide to Getting the Right Things Done***
    by Peter F. Drucker

22. ***Don't Make Me Think: A Common Sense Approach to Web Usability***
    by Steve Krug

23. ***The Deadline: A Novel About Project Management***
    by Tom DeMarco

24. ***The Psychology of Computer Programming***
    by Gerald M. Weinberg

25. ***Adrenaline Junkies and Template Zombies: Understanding Patterns of Project Behavior***
    by Tom Demarco, Peter Hruschka, Tim Lister, Steve McMenamin, Suzanne Robertson, and James Robertson

26. ***The Secrets of Consulting: A Guide to Giving and Getting Advice Successfully***
    by Gerald M. Weinberg

27. ***Death by Meeting***
    by Patrick Lencioni

28. ***The Advantage: Why Organizational Health Trumps Everything Else in Business***
    by Patrick Lencioni

29. ***Rise: 3 Practical Steps for Advancing Your Career, Standing Out as a Leader, and Liking Your Life***
    by Patty Azzarello

30. ***The Innovator's Solution: Creating and Sustaining Successful Growth***
    by Clayton M. Christensen and Michael E. Raynor

31. ***The Phoenix Project: A Novel about IT, DevOps, and Helping Your Business Win***
    by Gene Kim, Kevin Behr, and George Spafford

32. ***Accelerate: The Science of Lean Software and DevOps: Building and Scaling High Performing Technology Organizations***
    by Nicole Forsgren PhD, Jez Humble, and Gene Kim

7.3 Papers I've found very useful

I've long been a fan of hosting paper reading groups,[2] where a group of folks sit down and talk about interesting technical papers. One of the first steps to do that is identifying some papers worth chatting about, and here is a list of some papers I've seen lead to excellent discussions!

1. **"Dynamo: Amazon's Highly Available Key-Value Store"**

   If you read only the abstract, you'd be forgiven for not being overly excited about the Dynamo paper. This paper presents the design and implementation of Dynamo, a highly available key-value storage system that some of Amazon's core services use to provide an always-on experience. To achieve this level of availability, Dynamo sacrifices consistency under certain failure scenarios. It makes extensive use of object versioning and application-assisted conflict resolution in a manner that provides a novel interface for developers to use.

That said, this is in some senses "the" classic modern systems paper. It has happened more than once that an engineer I've met has only read a single systems paper in their career, and that paper was the Dynamo paper. This paper is a phenomenal introduction to eventual consistency, coordinating state across distributed storage, reconciling data as it diverges across replicas, and much more.

2. **"Hints for Computer System Design"**

   Butler Lampson[3] is winner of the ACM Turing Award (among other awards), and worked at the Xerox PARC. This paper concisely summarizes many of his ideas around systems design, and is a great read.

   In his words:

   > Studying the design and implementation of a number of computers has led to some general hints for system design. They are described here and illustrated by many examples, ranging from hardware such as the Alto and the Dorado to application programs such as Bravo and Star.

   This paper itself acknowledges that it doesn't aim to break any new ground, but it's a phenomenal overview.

3. **"Big Ball of Mud"**

   A reaction against exuberant publications about grandiose design patterns, this paper labels the most frequent architectural pattern as the Big Ball of Mud, and explores why elegant initial designs rarely remain intact as a system goes from concept to solution.

   From the abstract:

   > While much attention has been focused on high-level software architectural patterns, what is, in effect, the de-facto standard software architecture is seldom discussed. This paper examines this most frequently deployed of software architectures: the BIG BALL OF MUD. A BIG BALL OF MUD is a casually, even haphazardly, structured system. Its organization, if one can call it that, is dictated more by expe-

> diency than design. Yet, its enduring popularity cannot merely be indicative of a general disregard for architecture.

Although humor certainly infuses this paper, it's also correct that software design is remarkably poor. Very few systems have a design phase and few of those resembe the initial design (and documentation is rarely updated to reflect later decisions), making this an important topic for consideration.

4. **"The Google File System"**

From the abstract:

> The file system has successfully met our storage needs. It is widely deployed within Google as the storage platform for the generation and processing of data used by our service as well as research and development efforts that require large data sets. The largest cluster to date provides hundreds of terabytes of storage across thousands of disks on over a thousand machines, and it is concurrently accessed by hundreds of clients.
>
> In this paper, we present file system interface extensions designed to support distributed applications, discuss many aspects of our design, and report measurements from both micro-benchmarks and real-world use. Google has done something fairly remarkable in defining the technical themes in Silicon Valley and, at least debatably, across the entire technology industry. The company has been doing so for more than the last decade, and was only recently joined, to a lesser extent, by Facebook and Twitter as they reached significant scale. Google has defined these themes largely through noteworthy technical papers. The Google File System (GFS) paper is one of the early entries in that strategy, and is also remarkable as the paper that largely inspired the Hadoop File System (HFS).

5. **"On Designing and Deploying Internet-Scale Services"**

We don't always remember to consider Microsoft as one of the largest internet technology players—although, increasingly, Azure is making that comparison obvious and immediate—and it certainly wasn't a name that necessarily came to mind in 2007. This excellent paper from James Hamilton, which explores tips on building operable systems at extremely large scale, makes it clear that not considering Microsoft as a large internet player was a lapse in our collective judgment.

From the abstract:

> The system-to-administrator ratio is commonly used as a rough metric to understand administrative costs in high-scale services. With smaller, less automated services this ratio can be as low as 2:1, whereas on industry leading, highly automated services, we've seen ratios as high as 2,500:1. Within Microsoft services, Autopilot [1] is often cited as the magic behind the success of the Windows Live Search team in achieving high system-to-administrator ratios. While auto-administration is important, the most important factor is actually the service itself. Is the service efficient to automate? Is it what we refer to more generally as operations-friendly? Services that are operations-friendly require little human intervention, and both detect and recover from all but the most obscure failures without administrative intervention. This paper summarizes the best practices accumulated over many years in scaling some of the largest services at MSN and Windows Live.

This is a true checklist of how to design and evaluate large-scale systems (almost in the way that The Twelve-Factor App[4] wants to serve as a checklist for operable applications).

6. "CAP Twelve Years Later: How the 'Rules' Have Changed"

Eric Brewer posited the CAP theorem[5] in the early aughts, and 12 years later he wrote this excellent overview and review of CAP (which argues that distributed systems have to pick between either availability or consistency during partitions), Here's the rationale for the paper, in Brewer's words:

> In the decade since its introduction, designers and researchers have used (and sometimes abused) the CAP theorem as a reason to explore a wide variety of novel distributed systems. The NoSQL movement also has applied it as an argument against traditional databases.

CAP is interesting because there is not a "seminal CAP paper," but this article serves well in such a paper's stead. These ideas are expanded on in the "Harvest and Yield" paper.[6]

7. "Harvest, Yield, and Scalable Tolerant Systems"

This paper builds on the concepts from CAP Twelve Years Later, introducing the concepts of harvest and yield to add more nuance to the discussion about AP versus CP.

> The cost of reconciling consistency and state management with high availability is highly magnified by the unprecedented scale and robustness requirements of today's internet applications. We propose two strategies for improving overall availability using simple mechanisms that scale over large applications whose output behavior tolerates graceful degradation. We characterize this degradation in terms of harvest and yield, and map it directly onto engineering mechanisms that enhance availability by improving fault isolation, and in some cases also simplify programming. By collecting examples of related techniques in the literature and illustrating the surprising range of applications that can benefit from these approaches, we hope to motivate a broader research program in this area.

The harvest and yield concepts are particularly interesting because they are both self-evident and very rarely explicitly used; instead, distributed systems continue to fail in mostly undefined ways. Hopefully, as we keep rereading this paper, we'll also start to incorporate its design concepts into the systems we subsequently build!

8. "MapReduce: Simplified Data Processing on Large Clusters"

The MapReduce paper is an excellent example of an idea that has been so successful that it now seems self-evident. The idea of applying the concepts of functional programming at scale became a clarion call, provoking a shift from data warehousing to a new paradigm for data analysis:

> MapReduce is a programming model and an associated implementation for processing and generating large data sets. Users specify a map function that processes a key/value pair to generate a set of intermediate key/value pairs, and a reduce function that merges all intermediate values associated with the same intermediate key. Many real-world tasks are expressible in this model, as shown in the paper.

Much like the Google File System paper was an inspiration for the Hadoop File System, this paper was itself a major inspiration for Hadoop.

9. **"Dapper, a Large-Scale Distributed Systems Tracing Infrastructure"**

The Dapper paper introduces a performant approach to tracing requests across many services, which has become increasingly relevant as more companies refactor core monolithic applications into dozens or hundreds of micro-services.

From the abstract:

> Here we introduce the design of Dapper, Google's production distributed systems tracing infrastructure, and describe how our design goals of low overhead, application-level transparency, and ubiquitous deployment on a very large-scale system were met. Dapper shares conceptual similarities with other tracing systems, particularly Magpie and X-Trace, but certain design choices were made that have been key to its success in our environment, such as the use of sampling and restricting the instrumentation to a rather small number of common libraries.

The ideas from Dapper have since made their way into open source, especially in Zipkin[7] and OpenTracing.[8]

10. **"Kafka: a Distributed Messaging System for Log Processing"**

Apache Kafka[9] has become a core piece of infrastructure for many internet companies. Its versatility lends it to many roles, serving as the ingress point to "data land" for some and as a durable queue for others. And that's just scratching the surface.

Kafka is not only a useful addition to your tool kit, it's also a beautifully designed system:

> Log processing has become a critical component of the data pipeline for consumer internet companies. We introduce Kafka, a distributed messaging system that we developed for collecting and delivering high volumes of log data with low latency. Our system incorporates

> ideas from existing log aggregators and messaging systems, and is suitable for both offline and online message consumption. We made quite a few unconventional yet practical design choices in Kafka to make our system efficient and scalable. Our experimental results show that Kafka has superior performance when compared to two popular messaging systems. We have been using Kafka in production for some time and it is processing hundreds of gigabytes of new data each day.

In particular, Kafka's partitions do a phenomenal job of forcing application designers to make explicit decisions about trading off performance for predictable message ordering.

11. **"Wormhole: Reliable Pub-Sub to Support Geo-Replicated Internet Services"**

In many ways similar to Kafka, Facebook's Wormhole is another highly scalable approach to messaging:

> Wormhole is a publish-subscribe (pub-sub) system developed for use within Facebook's geographically replicated datacenters. It is used to reliably replicate changes among several Facebook services including TAO, Graph Search, and Memcache. This paper describes the design and implementation of Wormhole as well as the operational challenges of scaling the system to support the multiple data storage systems deployed at Facebook. Our production deployment of Wormhole transfers over 35 GBytes/sec in steady state (50 millions messages/sec or 5 trillion messages/day) across all deployments with bursts up to 200 GBytes/sec during failure recovery. We demonstrate that Wormhole publishes updates with low latency to subscribers that can fail or consume updates at varying rates, without compromising efficiency.

In particular, note the approach to supporting lagging consumers without sacrificing overall system throughput.

12. **"Borg, Omega, and Kubernetes"**

While the individual papers for each of Google's orchestration systems (Borg, Omega, and Kubernetes) are worth reading in their own right, this article is an excellent overview of the three:

> Though widespread interest in software containers is a relatively recent phenomenon, at Google we have been managing Linux containers at scale for more than ten years and built three different container-management systems in that time. Each system was heavily influenced by its predecessors, even though they were developed for different reasons. This article describes the lessons we've learned from developing and operating them.

Fortunately, not all orchestration happens under Google's aegis, and Apache Mesos's alternative two-layer scheduling architecture is a fascinating read as well.

13. "Large-Scale Cluster Management at Google with Borg"

Borg has orchestrated much of Google's infrastructure for quite some time (significantly predating Omega, although, fascinatingly, the Omega paper predates the Borg paper by two years):

> Google's Borg system is a cluster manager that runs hundreds of thousands of jobs, from many thousands of different applications, across a number of clusters each with up to tens of thousands of machines.

This paper takes a look at Borg's centralized scheduling model, which was both effective and efficient, although it became increasingly challenging to modify and scale over time. Borg inspired both Omega and Kubernetes within Google (the former to optimistically replace it, and the latter to seemingly commercialize the designers' learnings, or at least to prevent Mesos from capturing too much mind share).

14. "Omega: Flexible, Scalable Schedulers for Large Compute Clusters"

Omega is, among many other things, an excellent example of the second-system effect,[10] in which an attempt to replace a complex existing system with something far more elegant ends up being more challenging than anticipated.

In particular, Omega is a reaction against the realities of extending the aging Borg system:

> Increasing scale and the need for rapid response to changing requirements are hard to meet with current monolithic cluster scheduler architectures. This restricts the rate at which new features can be deployed, decreases efficiency and utilization, and will eventually limit cluster growth. We present a novel approach to address these needs using parallelism, shared state, and lock-free optimistic concurrency control.

Perhaps it's also an example of "worse is better"[11] once again taking the day.

15. "Mesos: A Platform for Fine-Grained Resource Sharing in the Data Center"

This paper describes the design of Apache Mesos,[12] in particular its distinctive two-level scheduler:

> We present Mesos, a platform for sharing commodity clusters between multiple diverse cluster computing frameworks, such as Hadoop and MPI. Sharing improves cluster utilization and avoids per-framework data replication. Mesos shares resources in a fine-grained manner, allowing frameworks to achieve data locality by taking turns reading data stored on each machine. To support the sophisticated schedulers of today's frameworks, Mesos introduces a distributed two-level scheduling mechanism called resource offers. Mesos decides how many resources to offer each framework, while frameworks decide which resources to accept and which computations to run on them.
>
> Our results show that Mesos can achieve near-optimal data locality when sharing the cluster among diverse frameworks, can scale to 50,000 (emulated) nodes, and is resilient to failures.

Used heavily by Twitter and Apple, Mesos was for some time the only open-source general scheduler with significant adoption. It's now in a fascinating competition for mind share with Kubernetes.

16. "Design Patterns for Container-Based Distributed Systems"

The move to container-based deployment and orchestration has introduced a whole new set of vocabulary, including "sidecars" and "adapters." This paper provides a survey of the patterns that have evolved over the past decade, as microservices and containers have become increasingly prominent infrastructure components:

> In the late 1980s and early 1990s, object-oriented programming revolutionized software development, popularizing the approach of building of applications as collections of modular components. Today we are seeing a similar revolution in distributed system development, with the increasing popularity of microservice architectures built from containerized software components. Containers are particularly well-suited as the fundamental "object" in distributed systems by virtue of the walls they erect at the container boundary. As this architectural style matures, we are seeing the emergence of design patterns, much as we did for object-oriented programs, and for the same reason—thinking in terms of objects (or containers) abstracts away the low-level details of code, eventually revealing higher-level patterns that are common to a variety of applications and algorithms.

The term "sidecar" in particular, likely originated in this blog post from Netflix,[13] which is a worthy read in its own right.

17. "Raft: In Search of an Understandable Consensus Algorithm"

We often see the second-system effect when a second system becomes bloated and complex relative to a simple initial system. However, the roles are reversed in the case of Paxos and Raft. Whereas Paxos is often considered beyond human comprehension, Raft is a fairly easy read:

> Raft is a consensus algorithm for managing a replicated log. It produces a result equivalent to (multi-)Paxos, and it is as efficient as Paxos, but its structure is different from Paxos; this makes Raft more understandable than Paxos and also provides a better foundation for building practical systems. In order to enhance understandability, Raft separates the key elements of consensus, such as leader election, log replication, and safety, and it enforces a stronger degree of coherency to reduce the number of states that must be considered.

> Results from a user study demonstrate that Raft is easier for students to learn than Paxos. Raft also includes a new mechanism for changing the cluster membership, which uses overlapping majorities to guarantee safety.

Raft is used by etcd[14] and influxDB[15] among many others.

18. **"Paxos Made Simple"**

One of Leslie Lamport's numerous influential papers, "Paxos Made Simple" is a gem, both in explaining the notoriously complex Paxos algorithm and because, even at its simplest, Paxos isn't really that simple:

> The Paxos algorithm for implementing a fault-tolerant distributed system has been regarded as difficult to understand, perhaps because the original presentation was Greek to many readers. In fact, it is among the simplest and most obvious of distributed algorithms. At its heart is a consensus algorithm—the "synod" algorithm. The next section shows that this consensus algorithm follows almost unavoidably from the properties we want it to satisfy. The last section explains the complete Paxos algorithm, which is obtained by the straightforward application of consensus to the state machine approach for building a distributed system—an approach that should be well-known, since it is the subject of what is probably the most often-cited article on the theory of distributed systems.

Paxos itself remains a deeply innovative concept, and is the algorithm behind Google's Chubby and Apache Zookeeper,[16] among many others.

19. **"SWIM: Scalable Weakly-Consistent Infection-Style Process Group Membership Protocol"**

The majority of consensus algorithms focus on being consistent during partition, but SWIM goes the other direction and focuses on availability:

> Several distributed peer-to-peer applications require weakly-consistent knowledge of process group membership information at all participating processes. SWIM is a generic software module that offers

> this service for large-scale process groups. The SWIM effort is motivated by the unscalability of traditional heart-beating protocols, which either impose network loads that grow quadratically with group size, or compromise response times or false positive frequency w.r.t. detecting process crashes. This paper reports on the design, implementation, and performance of the SWIM sub-system on a large cluster of commodity PCs.

SWIM is used in HashiCorp's software, as well as Uber's Ringpop.

20. **"The Byzantine Generals Problem"**

Another classic Leslie Lamport paper on consensus, the *Byzantine Generals Problem* explores how to deal with distributed actors that intentionally or accidentally submit incorrect messages:

> Reliable computer systems must handle malfunctioning components that give conflicting information to different parts of the system. This situation can be expressed abstractly in terms of a group of generals of the Byzantine army camped with their troops around an enemy city. Communicating only by messenger, the generals must agree upon a common battle plan. However, one or more of them may be traitors who will try to confuse the others. The problem is to find an algorithm to ensure that the loyal generals will reach agreement. It is shown that, using only oral messages, this problem is solvable if and only if more than two-thirds of the generals are loyal; so a single traitor can confound two loyal generals. With unforgeable written messages, the problem is solvable for any number of generals and possible traitors. Applications of the solutions to reliable computer systems are then discussed.

The paper is mostly focused on the formal proof, a bit of a theme from Lamport, who developed TLA+[17] to make formal proving easier, but it's also a useful reminder that we still tend to assume our components will behave reliably and honestly, and perhaps we shouldn't!

21. **"Out of the Tar Pit"**

"Out of the Tar Pit" bemoans unnecessary complexity in software, and proposes that functional programming and better data modeling can help us reduce accidental complexity. (The authors argue that most unnecessary complexity comes from state.)

From the abstract:

> Complexity is the single major difficulty in the successful development of large-scale software systems. Following Brooks we distinguish *accidental* from *essential* difficulty, but disagree with his premise that most complexity remaining in contemporary systems is essential. We identify common causes of complexity and discuss general approaches which can be taken to eliminate them where they are accidental in nature. To make things more concrete we then give an outline for a potential complexity-minimizing approach based on *functional programming* and *Codd's relational model of data*.

The paper's certainly a good read, although reading it a decade later, it's fascinating to see that neither of those approaches have particularly taken off. Instead the closest "universal" approach to reducing complexity seems to be the move to numerous mostly stateless services. This is perhaps more a reduction of local complexity, at the expense of larger systemic complexity, whose maintenance is then delegated to more specialized systems engineers.

(This is yet another paper that makes me wish TLA+ felt natural enough to be a commonly adopted tool.)

22. "The Chubby Lock Service for Loosely-Coupled Distributed Systems"

Distributed systems are hard enough without having to frequently reimplement Paxos or Raft. The model proposed by Chubby is to implement consensus once in a shared service, which will allow systems built upon it to share in the resilience of distribution by following greatly simplified patterns.

From the abstract:

> We describe our experiences with the Chubby lock service, which is intended to provide coarse-grained locking as well as reliable (though low-volume) storage for a loosely-coupled distributed system. Chubby provides an interface much like a distributed file sys-

> tem with advisory locks, but the design emphasis is on availability and reliability, as opposed to high performance. Many instances of the service have been used for over a year, with several of them each handling a few tens of thousands of clients concurrently. The paper describes the initial design and expected use, compares it with actual use, and explains how the design had to be modified to accommodate the differences.

In the open source world, the way Zookeeper is used in projects like Kafka and Mesos has the same role as Chubby.

23. **"Bigtable: A Distributed Storage System for Structured Data"**

One of Google's preeminent papers and technologies is Bigtable, which was an early (early in the internet era, anyway) NoSQL data store, operating at extremely high scale and built on top of Chubby.

> Bigtable is a distributed storage system for managing structured data that is designed to scale to a very large size: petabytes of data across thousands of commodity servers. Many projects at Google store data in Bigtable, including web indexing, Google Earth, and Google Finance. These applications place very different demands on Bigtable, both in terms of data size (from URLs to web pages to satellite imagery) and latency requirements (from backend bulk processing to real-time data serving). Despite these varied demands, Bigtable has successfully provided a flexible, high-performance solution for all of these Google products. In this paper we describe the simple data model provided by Bigtable, which gives clients dynamic control over data layout and format, and we describe the design and implementation of Bigtable.

From the SSTable design to the bloom filters, Cassandra inherits significantly from the Bigtable paper, and is probably rightfully considered a merging of the Dynamo and Bigtable papers.

24. **"Spanner: Google's Globally-Distributed Database"**

Many early NoSQL storage systems traded eventual consistency for increased resiliency, but building on top of eventually consistent systems can be harrowing. Spanner represents an approach from Google to offering both strong consistency and distributed reliability, based in part on a novel approach to managing time.

> Spanner is Google's scalable, multi-version, globally distributed, and synchronously-replicated database. It is the first system to distribute data at global scale and support externally-consistent distributed transactions. This paper describes how Spanner is structured, its feature set, the rationale underlying various design decisions, and a novel time API that exposes clock uncertainty. This API and its implementation are critical to supporting external consistency and a variety of powerful features: nonblocking reads in the past, lock-free read-only transactions, and atomic schema changes, across all of Spanner.

We haven't seen any open source Spanner equivalents yet, but I imagine we'll start seeing them soon.

25. **"Security Keys: Practical Cryptographic Second Factors for the Modern Web"**

Security keys like the YubiKey[18] have emerged as the most secure second authentication factor, and this paper out of Google explains the motivations that led to their creation, as well as the design that makes them work.

From the abstract:

> Security Keys are second-factor devices that protect users against phishing and man-in-the-middle attacks. Users carry a single device and can self-register it with any online service that supports the protocol. The devices are simple to implement and deploy, simple to use, privacy preserving, and secure against strong attackers. We have shipped support for Security Keys in the Chrome web browser and in Google's online services. We show that Security Keys lead to both an increased level of security and user satisfaction by analyzing a two-year deployment which began within Google and has extended to our consumer-facing web applications. The Security Key design has been standardized by the FIDO Alliance, an organization with more than 250 member companies spanning the industry. Currently, Security Keys have been deployed by Google, Dropbox, and GitHub.

These keys are also remarkably cheap! Order a few and start securing your life in a day or two.

26. "BeyondCorp: Design to Deployment at Google"

Building on the original BeyondCorp paper,[19] which was published in 2014, this paper is slightly more detailed and benefits from two more years of migration-fueled wisdom. That said, the big ideas have remained fairly consistent, and there is not much new relative to the BeyondCorp paper itself. If you haven't read that fantastic paper, this is an equally good starting point:

> The goal of Google's BeyondCorp initiative is to improve our security with regard to how employees and devices access internal applications. Unlike the conventional perimeter security model, BeyondCorp doesn't gate access to services and tools based on a user's physical location or the originating network; instead, access policies are based on information about a device, its state, and its associated user. BeyondCorp considers both internal networks and external networks to be completely untrusted, and gates access to applications by dynamically asserting and enforcing levels, or tiers, of access.

As is often the case when I read Google papers, my biggest takeaway here is to wonder when we'll start to see reusable, pluggable open source versions of the techniques described within.

27. "Availability in Globally Distributed Storage Systems"

This paper explores how to think about availability in replicated distributed systems, and is a useful starting point for those of us who are trying to determine the correct way to measure uptime for our storage layer or for any other sufficiently complex system.

From the abstract:

> We characterize the availability properties of cloud storage systems based on an extensive one-year study of Google's main storage infrastructure and present statistical models that enable further insight into the impact of multiple design choices, such as data placement and replication strategies. With these models we compare data avail-

> ability under a variety of system parameters given the real patterns of failures observed in our fleet.

Particularly interesting is the focus on correlated failures, building on the premise that users of distributed systems only experience the failure when multiple components have overlapping failures. Another expected but reassuring observation is that at Google's scale (and with resources distributed across racks and regions), most failure comes from tuning and system design, not from the underlying hardware.

I was also surprised by how simple their definition of availability was in this case:

> A storage node becomes *unavailable* when it fails to respond positively to periodic health checking pings sent by our monitoring system. The node remains unavailable until it regains responsiveness or the storage system reconstructs the data from other surviving nodes.

Often, discussions of availability become arbitrarily complex ("It should really be that response rates are over X, but with correct results and within our latency SLO!"), and it's reassuring to see the simplest definitions are still usable.

28. **"Still All on One Server: Perforce at Scale"**

As a company grows, code hosting performance becomes one of the critical factors in overall developer productivity (along with build and test performance), but it's a topic that isn't discussed frequently. This paper from Google discusses their experience scaling Perforce:

> Google runs the busiest single Perforce server on the planet, and one of the largest repositories in any source control system. From this high-water mark this paper looks at server performance and other issues of scale, with digressions into where we are, how we got here, and how we continue to stay one step ahead of our users.

This paper is particularly impressive when you consider the difficulties that companies run into as they scale Git monorepos (talk to an ex-Twitter employee near you for war stories).

29. "Large-Scale Automated Refactoring Using ClangMR"

Large codebases tend to age poorly, especially in the case of monorepos storing hundreds or thousands of different teams collaborating on different projects.

This paper covers one of Google's attempts to reduce the burden of maintaining their large monorepo through tooling that makes it easy to rewrite abstract syntax trees (ASTs) across the entire codebase.

From the abstract:

> In this paper, we present a real-world implementation of a system to refactor large C++ codebases efficiently. A combination of the Clang compiler framework and the MapReduce parallel processor, ClangMR enables code maintainers to easily and correctly transform large collections of code. We describe the motivation behind such a tool, its implementation and then present our experiences using it in a recent API update with Google's C++ codebase.

Similar work is being done with Pivot.[22]

30. "Source Code Rejuvenation is not Refactoring"

This paper introduces the concept of "code rejuvenation," a unidirectional process of moving toward cleaner abstractions as new language features and libraries become available, which is particularly applicable to sprawling, older codebases.

From the abstract:

> In this paper, we present the notion of *source code rejuvenation*, the automated migration of legacy code and very briefly mention the tools we use to achieve that. While *refactoring* improves structurally inadequate source code, source code rejuvenation leverages enhanced program language and library facilities by finding and replacing coding patterns that can be expressed through higher-level software abstractions. Raising the level of abstraction benefits software maintainability, security, and performance.

There are some strong echoes of this work in Google's ClangMR paper.[20]

31. "Searching for Build Debt: Experiences Managing Technical Debt at Google"

This paper is an interesting cover of how to perform large-scale migrations in living codebases. Using broken builds as the running example, they break down their strategy into three pillars: automating, making it easy to do the right thing, and making it hard to do the wrong thing.

From the abstract:

> With a large and rapidly changing codebase, Google software engineers are constantly paying interest on various forms of technical debt. Google engineers also make efforts to pay down that debt, whether through special Fixit days, or via dedicated teams, variously known as janitors, cultivators, or demolition experts. We describe several related efforts to measure and pay down technical debt found in Google's BUILD files and associated dead code. We address debt found in dependency specifications, unbuildable targets, and unnecessary command line flags. These efforts often expose other forms of technical debt that must first be managed.

32. "No Silver Bullet—Essence and Accident in Software Engineering"

A seminal paper from the author of *The Mythical Man-Month,* "No Silver Bullet" expands on discussions of accidental versus essential complexity, and argues that there is no longer enough accidental complexity to allow individual reductions in that accidental complexity to significantly increase engineer productivity.

From the abstract:

> Most of the big past gains in software productivity have come from removing artificial barriers that have made the accidental tasks inordinately hard, such as severe hardware constraints, awkward programming languages, lack of machine time. How much of what software engineers now do is still devoted to the accidental, as

> opposed to the essential? Unless it is more than 9/10 of all effort, shrinking all the accidental activities to zero time will not give an order of magnitude improvement.

I think that, interestingly, we do see accidental complexity in large codebases become large enough to make order-of-magnitude improvements (motivating, for example, Google's investments in ClangMR and such), so perhaps we're not quite as far ahead in the shift to essential complexity as we'd like to believe.

33. "The UNIX Time-Sharing System"

This paper describes the fundamentals of UNIX as of 1974. What is truly remarkable is how many of the design decisions are still used today. From the permission model that we've all manipulated with chmod to system calls used to manipulate files, it's amazing how much remains intact.

From the abstract:

> UNIX is a general-purpose, multi-user, interactive operating system for the Digital Equipment Corporation PDP-11/40 and 11/45 computers. It offers a number of features seldom found even in larger operating systems, including: (1) a hierarchical file system incorporating demountable volumes; (2) compatible file, device, and interprocess I/O; (3) the ability to initiate asynchronous processes; (4) system command language selectable on a per-user basis; and (5) over 100 subsystems including a dozen languages. This paper discusses the nature and implementation of the file system and of the user command interface.

Also fascinating is their observation that UNIX has in part succeeded because it was designed to solve a general problem by its authors (working with the PDP-7 was frustrating), rather than to progress toward a more specified goal.

Endnotes

Chapter 2: Organizations

1. https://lethain.com/running-an-engineering-reorg/
2. https://lethain.com/strategies-visions/
3. https://lethain.com/guiding-broad-change-with-metrics/
4. https://increment.com/on-call/
5. https://lethain.com/durably-excellent-teams/
6. https://lethain.com/running-an-engineering-reorg/
7. http://www.askmar.com/Business%20Plan/2006-11-18%20Yahoo%20Peanut%20Butter%20Manifesto.pdf
8. https://lethain.com/durably-excellent-teams/
9. https://lethain.com/durably-excellent-teams/
10. https://www.amazon.com/dp/B002LHRM2O
11. https://www.amazon.com/Thinking-Systems-Donella-H-Meadows/dp/1603580557
12. https://hbr.org/2008/04/managing-hypergrowth

13. http://fortune.com/2015/03/15/bill-gurley-predicts-dead-unicorns-in-startup-land-this-year/

14. https://www.amazon.com/Mythical-Man-Month-Software-Engineering-Anniversary/dp/0201835959

15. https://www.amazon.com/dp/B00A-ZRBLHO/

16. https://lethain.com/tools-for-operating-a-growing-org/

17. https://lethain.com/identify-your-controls/

18. https://lethain.com/productivity-in-the-age-of-hypergrowth/

Chapter 3: Tools

1. https://lethain.com/building-technical-leverage/

2. https://lethain.com/intro-product-management/

3. https://www.amazon.com/Thinking-Systems-Donella-H-Meadows/dp/1603580557

4. https://lethain.com/accelerate-developer-productivity/

5. https://www.iseesystems.com/

6. https://insightmaker.com/

7. https://lethain.com/durably-excellent-teams/

8. https://lethain.com/product-management-infra-engineering/

9. https://www.amazon.com/dp/B004W3FM4A/ref=dp-kindle-redirect?_encoding=UTF8andbtkr=1

10. https://lethain.com/building-technical-leverage/

11. https://lethain.com/migrations/

12. https://lethain.com/migrations/

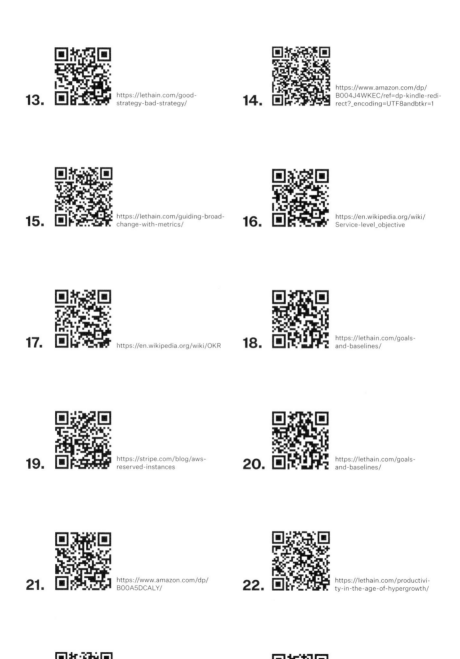

13. https://lethain.com/good-strategy-bad-strategy/

14. https://www.amazon.com/dp/B004J4WKEC/ref=dp-kindle-redirect?_encoding=UTF8andbtkr=1

15. https://lethain.com/guiding-broad-change-with-metrics/

16. https://en.wikipedia.org/wiki/Service-level_objective

17. https://en.wikipedia.org/wiki/OKR

18. https://lethain.com/goals-and-baselines/

19. https://stripe.com/blog/aws-reserved-instances

20. https://lethain.com/goals-and-baselines/

21. https://www.amazon.com/dp/B00A5DCALY/

22. https://lethain.com/productivity-in-the-age-of-hypergrowth/

23. https://en.wikipedia.org/wiki/You_aren%27t_gonna_need_it

24. https://lethain.com/refactoring-programmatically/

25. https://en.wikipedia.org/wiki/Lint_(software)

26. https://lethain.com/goals-and-baselines/

27. https://lethain.com/strategies-visions/

28. https://lethain.com/strategies-visions/

29. https://lethain.com/roles-over-rocket-ships/

30. https://lethain.com/partnering-with-your-manager/

31. https://lethain.com/digg-v4-architecture-process/

32. https://www.amazon.com/ALL-NEW-Dont-Think-Elephant/dp/160358594X/

33. https://en.wikipedia.org/wiki/Kanban

34. https://en.wikipedia.org/wiki/Lint_(software)

35. https://lethain.com/selecting-project-leads/

36. https://en.wikipedia.org/wiki/Yahoo!_Search_BOSS

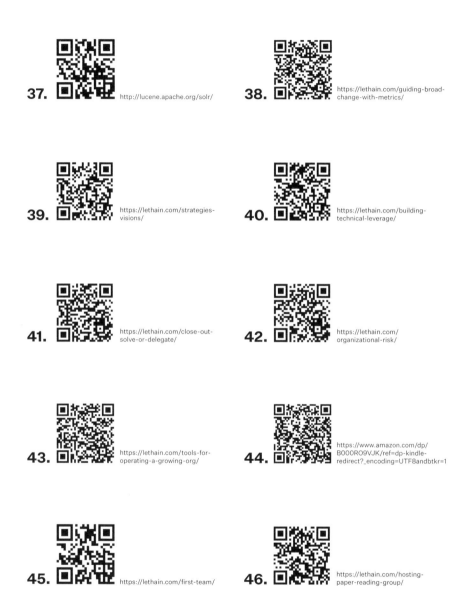

37. http://lucene.apache.org/solr/

38. https://lethain.com/guiding-broad-change-with-metrics/

39. https://lethain.com/strategies-visions/

40. https://lethain.com/building-technical-leverage/

41. https://lethain.com/close-out-solve-or-delegate/

42. https://lethain.com/organizational-risk/

43. https://lethain.com/tools-for-operating-a-growing-org/

44. https://www.amazon.com/dp/B000RO9VJK/ref=dp-kindle-redirect?_encoding=UTF8andbtkr=1

45. https://lethain.com/first-team/

46. https://lethain.com/hosting-paper-reading-group/

Chapter 4: Approaches

1. https://lethain.com/strategies-visions/
2. https://www.amazon.com/Phoenix-Project-DevOps-Helping-Business/dp/0988262592
3. https://lethain.com/case-against-top-down-global-optimization/
4. https://lethain.com/productivity-in-the-age-of-hypergrowth/
5. https://lethain.com/infrastructure-between-cost-center-and-before-ego-trip/
6. https://en.wikipedia.org/wiki/Golden_Rule
7. https://medium.com/@evnowand-forever/f-you-i-quit-hiring-is-broken-bb8f3a48d324
8. https://en.wikipedia.org/wiki/Dialectic#Hegelian_dialectic
9. https://lethain.com/adding-value-as-an-engineering-manager/
10. https://lethain.com/ways-engineering-managers-get-stuck/
11. https://en.wikipedia.org/wiki/The_Good_Earth
12. https://lethain.com/ways-engineering-managers-get-stuck/

13. https://www.amazon.com/Getting-Right-Tao-Contemporary-Ching/dp/0982473982

14. https://lethain.com/some-of-my-favorite-technical-papers/

15. https://lethain.com/good-strategy-bad-strategy/

16. https://en.wikipedia.org/wiki/Sensitivity_analysis

Chapter 5: Culture

1. https://lethain.com/selecting-project-leads/

2. https://lethain.com/selecting-project-leads/

3. https://lethain.com/hosting-paper-reading-group/

4. https://www.donut.com/

5. https://lethain.com/sizing-engineering-teams/

6. https://www.attack-gecko.net/2018/06/25/building-a-first-team-mindset/

7. https://en.wikipedia.org/wiki/Rooney_Rule

8. https://en.wikipedia.org/wiki/Me_and_Bobby_McGee

9. http://classics.mit.edu/Plato/apology.html

10. https://plato.stanford.edu/entries/liberty-positive-negative/

11. https://plato.stanford.edu/entries/liberty-positive-negative/#ParPosLib

12. https://en.wikipedia.org/wiki/System_dynamics

13. https://www.amazon.com/Slack-Getting-Burnout-Busywork-Efficiency/dp/0767907698

14. https://a16z.com/2014/02/06/6147/

15. https://www.amazon.com/Innovators-Dilemma-Revolutionary-Change-Business/dp/0062060244

16. https://www.amazon.com/Thinking-Systems-Donella-H-Meadows/dp/1603580557/

17. https://en.wikipedia.org/wiki/Dust_Bowl

18. https://en.wikipedia.org/wiki/Foundation_series

Chapter 6: Careers

1. https://www.wired.com/story/surviving-as-an-old-in-the-tech-world/
2. https://blog.wealthfront.com/how-long-should-you-stay-at-your-job/
3. https://www.linkedin.com/in/will-larson-a44b543/
4. http://randsinrepose.com/archives/wanted/
5. https://www.comparably.com/blog/best-places-to-work-competition/
6. https://www.lever.co/
7. http://www.greenhouse.io/
8. https://lethain.com/membership-opportunity/
9. https://lethain.com/membership-opportunity/
10. https://jvns.ca/blog/2017/12/31/2017--year-in-review/
11. https://lethain.com/first-team/
12. https://www.businessinsider.com/stack-ranking-employees-is-a-bad-idea-2013-11

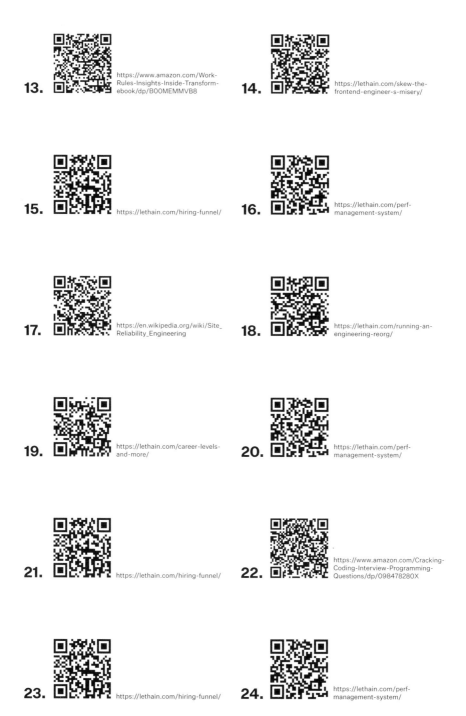

13. https://www.amazon.com/Work-Rules-Insights-Inside-Transform-ebook/dp/B00MEMMVB8

14. https://lethain.com/skew-the-frontend-engineer-s-misery/

15. https://lethain.com/hiring-funnel/

16. https://lethain.com/perf-management-system/

17. https://en.wikipedia.org/wiki/Site_Reliability_Engineering

18. https://lethain.com/running-an-engineering-reorg/

19. https://lethain.com/career-levels-and-more/

20. https://lethain.com/perf-management-system/

21. https://lethain.com/hiring-funnel/

22. https://www.amazon.com/Cracking-Coding-Interview-Programming-Questions/dp/098478280X

23. https://lethain.com/hiring-funnel/

24. https://lethain.com/perf-management-system/

25. https://lethain.com/perf-management-system/

Chapter 7: Appendix

1. https://twitter.com/davinbogan

2. https://lethain.com/hosting-paper-reading-group/

3. https://en.m.wikipedia.org/wiki/Butler_Lampson

4. https://12factor.net

5. https://en.wikipedia.org/wiki/CAP_theorem

6. https://www.semanticscholar.org/paper/Harvest%2C-Yield-and-Scalable-Tolerant-Systems-Fox-Brewer/50158bc1a8a67295ab7b-ce0550886a9859000dc2

7. https://zipkin.io/

8. https://opentracing.io/

9. http://kafka.apache.org/

10. http://catb.org/jargon/html/S/second-system-effect.html

11. https://www.jwz.org/doc/worse-is-better.html

12. http://mesos.apache.org/

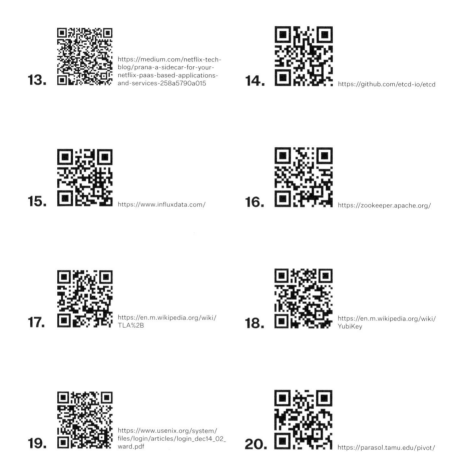

13. https://medium.com/netflix-tech-blog/prana-a-sidecar-for-your-netflix-paas-based-applications-and-services-258a5790a015

14. https://github.com/etcd-io/etcd

15. https://www.influxdata.com/

16. https://zookeeper.apache.org/

17. https://en.m.wikipedia.org/wiki/TLA%2B

18. https://en.m.wikipedia.org/wiki/YubiKey

19. https://www.usenix.org/system/files/login/articles/login_dec14_02_ward.pdf

20. https://parasol.tamu.edu/pivot/

21. https://ai.google/research/pubs/pub41342

Chapter 7: Appendix; Books I've found very useful

1. https://www.amazon.com/Thinking-Systems-Donella-H-Meadows/dp/1603580557/ref=sr_1_1?keywords=Thinking+in+Systems%3A+A+Primer+by+Donella+Meadows&qid=1551585388&s=gateway&sr=8-1

2. https://www.amazon.com/ALL-NEW-Dont-Think-Elephant/dp/160358594X/ref=sr_1_fkmrnull_1?keywords=Don't+Think+of+an+Elephant%21+Know+Your+Values+and+Frame+the+Debate+by+George+Lakoff&qid=1551585650&s=gateway&sr=8-1-fkmrnull

3. https://www.amazon.com/People-ware-Productive-Projects-Teams-3rd/dp/0321934113/ref=sr_1_fkmrnull_1?keywords=Peopleware%3A+Productive+Projects+and+Teams+by+DeMarco+and+Lister&qid=1551585745&s=gateway&sr=8-1-fkmrnull

4. https://www.amazon.com/Slack-Getting-Burnout-Busywork-Efficiency/dp/0767907698/ref=sr_1_fkmrnull_1?keywords=Slack%3A+Getting+Past+Burnout%2C+Busywork%2C+and+the+Myth+of+Total+Efficiency+by+Tom+DeMarco&qid=1551585846&s=gateway&sr=8-1-fkmrnull

5. https://www.amazon.com/Mythical-Man-Month-Software-Engineering-Anniversary/dp/0201835959/ref=sr_1_fkmrnull_1?keywords=The+Mythical+Man-Month+by+Frederick+Brooks&qid=1551585913&s=gateway&sr=8-1-fkmrnull

6. https://www.amazon.com/Good-Strategy-Bad-Difference-Matters-ebook/dp/B004J4WKEC/ref=sr_1_fk-mrnull_1?keywords=Good+Strategy+Bad+Strategy%3A+The+Difference+and+Why+it+Matters+by+Richard+Rumelt&qid=1551585977&s=gateway&sr=8-1-fkmrnull

7. https://www.amazon.com/Goal-Process-Ongoing-Improvement/dp/0884271951/ref=sr_1_1?keywords=The+Goal%3A+A+Process+of+Ongoing+Improvement+by+Eliyahu+Goldratt&qid=1551586027&s=gateway&sr=8-1

8. https://www.amazon.com/Five-Dysfunctions-Team-Leadership-Fable/dp/0787960756/ref=sr_1_1?keywords=The+Five+Dysfunctions+of+a+Team+by+Patrick+Lencioni&qid=1551586085&s=gateway&sr=8-1

9. https://www.amazon.com/Three-Signs-Miserable-Job/dp/8126552697/ref=sr_1_fkmrnull_1?keywords=The+Three+Signs+of+a+Miserable+Job+by+Patrick+Lencioni&qid=1551586178&s=gateway&sr=8-1-fkmrnull

10. https://www.amazon.com/Finite-Infinite-Games-James-Carse/dp/1476731713/ref=sr_1_1?keywords=Finite+and+Infinite+Games+by+James+Carse&qid=1551586314&s=gateway&sr=8-1

11. https://www.amazon.com/INSPIRED-Create-Tech-Products-Customers/dp/1119387507/ref=sr_1_fkmrnull_1?keywords=INSPIRED%3A+How+to+Create+Tech+Products+Customers+Love+by+Marty+Cagan&qid=1551586391&s=gateway&sr=8-1-fkmrnull

12. https://www.amazon.com/Innovators-Dilemma-Technologies-Management-Innovation/dp/1633691780/ref=sr_1_fkmrnull_1?keywords=The+Innovator's+Dilemma%3A+When+New+Technologies+Cause+Great+Firms+to+Fail+by+Clayton+Christensen&qid=1551586451&s=gateway&sr=8-1-fkmrnull

13. https://www.amazon.com/Myth-Revisited-Small-Businesses-About-ebook/dp/B000RO9VJK/ref=sr_1_fkmrnull_1?keywords=The+E-Myth+Revisited%3A+Why+Most+Small+Businesses+Don't+Work+and+What+to+Do+About+It+by+Michael+Gerber&qid=1551586507&s=gateway&sr=8-1-fkmrnull

14. https://www.amazon.com/Fierce-Conversations-Achieving-Success-Conversation/dp/B06XGNMDBY/ref=sr_1_fkmrnull_1?keywords=Fierce+Conversations%3A+Achieving+Success+at+Work+and+in+Life+One+Conversation+at+a+Time+by+Susan+Scott&qid=1551586587&s=gateway&sr=8-1-fkmrnull

15. https://www.amazon.com/Becoming-Technical-Leader-Problem-solving-Paperback/dp/B00NPOCUKQ/ref=sr_1_fkmr0_1?keywords=Becoming+a+Technical+Leader%3A+An+Organic+Problem-Solving+Approach+by+Gerald+Weinberg&qid=1551587007&s=books&sr=8-1-fkmr0

16. https://www.amazon.com/Designing-Mind-Understanding-Interface-Guidelines/dp/0124079148/ref=sr_1_fkmrnull_1?keywords=Designing+with+the+Mind+in+Mind+by+Jeff+Johnson&qid=1551587071&s=gateway&sr=8-1-fkmrnull

17. https://www.amazon.com/Leadership-Pipeline-Build-Powered-Company/dp/0470894563/ref=sr_1_fkmrnull_1?keywords=The+Leadership+Pipeline%3A+How+to+Build+the+Leadership+Powered+Company+by+Charan%2C+Drotter%2C+and+Noel&qid=1551587127&s=gateway&sr=8-1-fkmrnull

18. https://www.amazon.com/Managers-Path-Leaders-Navigating-Growth/dp/1491973897/ref=sr_1_fkmrnull_1?keywords=The+Manager's+Path%3A+A+Guide+for+Tech+Leaders+Navigating+Growth+and+Change+by+Camille+Fournier&qid=1551591945&s=gateway&sr=8-1-fkmrnull

19. https://www.amazon.com/High-Output-Management-Andrew-Grove/dp/0679762884/ref=sr_1_1?keywords=High+Output+Management+by+Andy+Grove&qid=1551592041&s=gateway&sr=8-1

20. https://www.amazon.com/First-90-Days-Strategies-Expanded/dp/1422188612/ref=sr_1_fkmrnull_1?keywords=The+First+90+Days%3A+Proven+Strategies+for+Getting+Up+to+Speed+Faster+and+Smarter%2C+Updated+and+Expanded+by+Michael+Watkins&qid=1551592130&s=gateway&sr=8-1-fkmrnull

21. https://www.amazon.com/Effective-Executive-Definitive-Harperbusiness-Essentials-ebook/dp/B01F1WZGNC/ref=sr_1_fkmrnull_1?keywords=The+Effective+Executive%3A+The+Definitive+Guide+to+Getting+the+Right+Things+Done+by+Peter+Drucker&qid=1551592182&s=gateway&sr=8-1-fkmrnull

22. https://www.amazon.com/Dont-Make-Me-Think-Usability/dp/0321344758/ref=sr_1_fkmrnull_2?keywords=Don't+Make+Me+Think%3A+A+Common+Sense+Approach+to+Web+Usability+by+Steve+Krug&qid=1551592244&s=gateway&sr=8-2-fkmrnull

23. https://www.amazon.com/Deadline-Novel-About-Project-Management/dp/0932633390/ref=sr_1_fkmrnull_1?keywords=The+Deadline%3A+A+Novel+About+Project+-Management+by+Tom+DeMarco&qid=1551592302&s=gateway&sr=8-1-fkmrnull

24. https://www.amazon.com/Psychology-Computer-Programming-Silver-Anniversary/dp/0932633420/ref=sr_1_fkmrnull_1?keywords=The+-Psychology+of+Computer+Programming+by+Gerald+Weinberg&qid=1551592351&s=gateway&sr=8-1-fkmrnull

25. https://www.amazon.com/Adrenaline-Junkies-Template-Zombies-Understanding-ebook/dp/B00DY3KQHM/ref=sr_1_fkmrnull_1?keywords=Adrenaline+Junkies+and+Template+Zombies%3A+Understanding+Patterns+of+Project+Behavior+by+DeMarco%2C+Hruschka%2C+Lister%2C+McMenamin%2C+Robertson%2C+and+Robertson&qid=1551592413&s=gateway&sr=8-1-fkmrnull

26. https://www.amazon.com/Secrets-Consulting-Giving-Getting-Successfully-ebook/dp/B004J35LHQ/ref=sr_1_fkmrnull_1?keywords=The+Secrets+of+Consulting%3A+A+Guide+to+Giving+and+Getting+Advice+Successfully+by+Gerald+Weinberg&qid=1551592463&s=gateway&sr=8-1-fkmrnull

27. https://www.amazon.com/Death-Meeting-Leadership-Solving-Business/dp/0787968056/ref=sr_1_1?keywords=Death+by+Meeting+by+Patrick+Lencioni&qid=1551592514&s=gateway&sr=8-1

28. https://www.amazon.com/Advantage-Organizational-Health-Everything-Business/dp/B007MIWCAY/ref=sr_1_fkmrnull_1?keywords=The+Advantage%3A+Why+Organizational+Health+Trumps+Everything+Else+in+Business+by+Patrick+Lencioni&qid=1551592591&s=gateway&sr=8-1-fkmrnull

29. https://www.amazon.com/Rise-Practical-Advancing-Career-Standing/dp/B0762349WT/ref=sr_1_1?keywords=Rise%3A+3+Practical+Steps+for+Advancing+Your+Career%2C+Standing+Out+as+a+Leader%2C+and+Liking+Your+Life+by+Patty+Azzarello&qid=1551592651&s=books&sr=1-1-catcorr

30. https://www.amazon.com/Innovators-Solution-Creating-Sustaining-Successful/dp/B000MGTPY4/ref=sr_1_1?keywords=The+Innovator's+Solution%3A+Creating+and+Sustaining+Successful+Growth+by+Christensen+and+Raynor&qid=1551592710&s=books&sr=1-1-catcorr

31. https://www.amazon.com/Phoenix-Project-Helping-Business-Anniversary/dp/B00VATFAMI/ref=sr_1_1?keywords=The+Phoenix+Project%3A+A+Novel+about+IT%2C+DevOps%2C+and+Helping+Your+Business+Win+by+Kim%2C+Behr%2C+and+Spafford&qid=1551592774&s=books&sr=1-1-catcorr

32. https://www.amazon.com/Nicole-Forsgren-Accelerate-Organizations-2018 /dp/B07J63DQB2/ref=sr_1_fkmr0_2?keywords=Accelerate%3A+The+Science+of+Lean+Software+and+DevOps%3A+Building+and+Scaling+High+Performing+Technology+Organizations+by+Forsgren%2C+Humble%2C+and+Kim&qid=1551592826&s=books&sr=8-2-fkmr0

Chapter 7: Appendix; Papers I've found very useful

1. https://www.allthingsdistributed.com/files/amazon-dynamo-sosp2007.pdf
2. https://www.microsoft.com/en-us/research/wp-content/uploads/2016/02/acrobat-17.pdf
3. https://www.researchgate.net/publication/2938621_Big_Ball_of_Mud
4. https://static.googleusercontent.com/media/research.google.com/en//archive/gfs-sosp2003.pdf
5. http://highscalability.com/paper-designing-and-deploying-internet-scale-services
6. https://www.researchgate.net/publication/220476881_CAP_Twelve_years_later_How_the_Rules_have_Changed
7. http://citeseerx.ist.psu.edu/viewdoc/download?doi=10.1.1.8.671&rep=rep1&type=pdf
8. https://static.googleusercontent.com/media/research.google.com/en//archive/mapreduce-osdi04.pdf
9. https://static.googleusercontent.com/media/research.google.com/en//archive/papers/dapper-2010-1.pdf
10. http://notes.stephenholiday.com/Kafka.pdf
11. https://www.usenix.org/system/files/conference/nsdi15/nsdi15-paper-sharma.pdf
12. https://static.googleusercontent.com/media/research.google.com/en//pubs/archive/44843.pdf

13. https://pdos.csail.mit.edu/6.824/papers/borg.pdf

14. https://static.googleusercontent.com/media/research.google.com/en//pubs/archive/41684.pdf

15. https://people.eecs.berkeley.edu/~alig/papers/mesos.pdf

16. https://static.googleusercontent.com/media/research.google.com/en//pubs/archive/45406.pdf

17. https://raft.github.io/raft.pdf

18. https://lamport.azurewebsites.net/pubs/paxos-simple.pdf

19. https://www.cs.cornell.edu/projects/Quicksilver/public_pdfs/SWIM.pdf

20. https://people.eecs.berkeley.edu/~luca/cs174/byzantine.pdf

21. https://github.com/papers-we-love/papers-we-love/blob/master/design/out-of-the-tar-pit.pdf

22. https://static.googleusercontent.com/media/research.google.com/en//archive/chubby-osdi06.pdf

23. https://static.googleusercontent.com/media/research.google.com/en//archive/bigtable-osdi06.pdf

24. https://static.googleusercontent.com/media/research.google.com/en//archive/spanner-osdi2012.pdf

25. https://ai.google/research/pubs/pub45409

26. https://ai.google/research/pubs/pub44860

27. https://static.googleusercontent.com/media/research.google.com/en//pubs/archive/36737.pdf

28. http://info.perforce.com/rs/perforce/images/GoogleWhitePaper-StillAllonOneServer-PerforceatScale.pdf

29. https://ai.google/research/pubs/pub41342

30. http://www.stroustrup.com/sofsem10.pdf

31. https://static.googleusercontent.com/media/research.google.com/en//pubs/archive/37755.pdf

32. http://worrydream.com/refs/Brooks-NoSilverBullet.pdf

33. https://people.eecs.berkeley.edu/~brewer/cs262/unix.pdf

Notes